Fleece
fantastic

Fleece
fantastic

35 cute, cozy, and quick projects to make and give

Rachel Henderson

CICO BOOKS
LONDON NEW YORK

This book is dedicated to my mum and dad, who bought me my first sewing machine and helped open up a whole new exciting world for me.

Published in 2014 by CICO Books
An imprint of Ryland Peters & Small
519 Broadway, 5th Floor, New York NY 10012
20–21 Jockey's Fields, London WC1R 4BW
www.rylandpeters.com

10 9 8 7 6 5 4 3 2 1

Text copyright © Rachel Henderson 2014
Design, photography, and illustration
copyright © CICO Books 2014

A CIP catalog record for this book is available from
the Library of Congress and the British Library.

ISBN: 978-1-78249-146-0

Printed in China

Editor: Sarah Hoggett
Designer: Alison Fenton
Photographer: Penny Wincer
Stylist: Catherine Woram and Isabel de Cordova
Illustrator: Michael A Hill
Template Illustrator: Kuo Kang Chen

In-House Editor: Carmel Edmonds
Publishing Manager: Penny Craig
Art Director: Sally Powell
Production Controller: David Hearn
Publisher: Cindy Richards

Contents

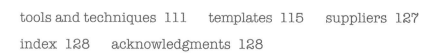

Introduction

My love affair with fleece came about when I started making accessories for children and realized how super-easy, quick, and fun it was to create handmade items. Unlike cotton fabrics, fleece doesn't fray at the edges, so some projects can be made with very little sewing or sometimes no sewing at all, and it's also great for appliqué.

The best thing about fleece is that it comes in such a wide range of bright, bold colors and decorative patterns, and it's also very cheap to buy, all of which make it the perfect fabric choice for beginners.

I wanted to write a book that would open your eyes to the possibilities of fleece, with the hope that it makes you look at it in a whole new light! There are so many stylish and contemporary designs you can make with this wonderfully versatile fabric. You can also work with it in lots of exciting ways—knotting it, braiding (plaiting) it, weaving it, folding it, and free-motion stitching onto it! I've tried to make some designs very simple and also included some more challenging ones, so that there's something for all sewing abilities.

Whether it's cheeky frog slippers, cozy boot warmers, or colorful fleecy coasters, every project is a joy to make—and whoever you make it for will, I'm sure, thoroughly enjoy wearing or using it.

Have fun!

Rachel

Chapter 1

Fashion accessories

In this chapter, there's a whole range of cozy projects for both indoor and outdoor wear. Most of the designs are very easy and quick to make, although there are a few that require a little more time and patience. You'll notice that I've played around with fleece quite a bit in this chapter— folding it, weaving it, and embellishing with free-motion machine embroidery. I hope this will give you the confidence to experiment and perhaps come up with your own designs!

This beautiful flower brooch is super quick to make, and a lovely accessory to attach to a bag, jacket, or cardigan.

flower brooch

Skill level: 1

You will need
Templates on page 115
10-in. (25-cm) square of
 fleece in your chosen
 color
Scrap of felt to match
 the fleece
Sewing thread to match
 the fleece
Brooch back
Sewing machine
Basic sewing kit (see
 page 111)

1 Trace the outer and inner petal templates on page 115 onto card (see page 112) and cut out. Place them on the wrong side of your chosen color of fleece, draw around each one five times with a permanent marker pen, and cut out all the petals.

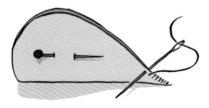

2 Place each inner petal right side up on top of an outer petal, with the points together. Fold each one in half lengthwise, so that the right sides are facing in, and pin to hold. Using matching thread and starting from the point, whipstitch (see page 113) the side edges together for ¼ in. (5 mm).

3 Overlap one petal on the next at the base, pin together, then whipstitch from the base upward for about ⅜ in. (1 cm). Repeat to attach the remaining petals.

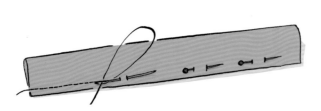

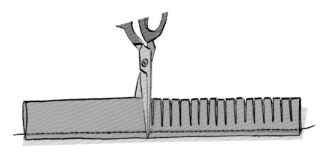

4 Cut a 6 x 1½-in. (15 x 4-cm) strip of fleece. Fold it in half lengthwise, so that the wrong sides are facing in, and pin to hold. Work a small backstitch (see page 112) by hand along the open edge (or straight stitch by machine).

5 Cut slits ⅝ in. (1.5 cm) long and ¼ in. (5 mm) apart along the folded edge to form small loops.

6 Place a small amount of glue along the stitched edge and roll the strip up tightly to form the flower center decoration.

7 Place the coiled strip in the center of the flower and attach using a small amount of fabric glue.

8 Turn each flower wrong side up. Draw a 2½-in. (6-cm) diameter circle on card and cut out. This is the backing template. Place it on matching felt, draw around it, and cut out.

9 Cut out a ⅜ x ¾-in. (1 x 2-cm) rectangle of felt and place it over the brooch back bar. Place the felt backing circle in the center of the flower back, with the brooch pin and rectangle of felt on top, and attach using fabric glue. Alternatively, hand stitch the brooch bar in place.

Don't let your feet get cold during those long country walks! Stitch up these stylish yet easy-to-make polka-dot boot warmers, featuring pretty bows, and stay nice and cozy.

spotty boot warmers

Skill level: 3

You will need

Templates on pages 118–119

2 yd (2 m) red polka-dot fleece

Red polyester sewing thread

Sewing machine

Basic sewing kit (see page 111)

Adhesive tape

To fit ladies' shoe sizes

Small: US 5.5–6.5 (UK 3–4)

Medium: US 7.5–8.5 (UK 5–6)

Large: US 9.5–11.5 (UK 7–9)

Making the boot warmers

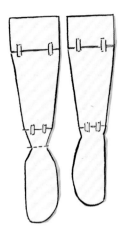

I   Transfer the templates on pages 118–119 onto card (see page 112) and cut out. Tape parts 1, 2, and 3 together to make the templates for the front and back.

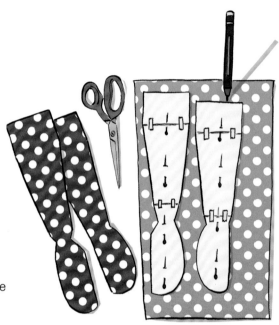

2   Place the front and back templates on the wrong side of the polka-dot fleece across the grainline, draw around them using a permanent marker pen, and cut out to make the left front and back. Flip the pattern pieces over. Still working on the wrong side of the fleece, draw around the patterns again, then cut out to make the right front and back.

3   With the right sides together, using the dotted line on the back pattern as a guide, fold over the heels. Pin in place and straight stitch around the heel by machine, creating a curved shape.

4   Pin the front and back pieces right sides together. Taking a ⅜-in. (1-cm) seam allowance and leaving the top edge open, machine stitch the pieces together. Turn right side out.

Making the cuffs

5 For the cuff, draw a rectangle 5 x 16 in./13 x 41 cm (small); 5 x 17 in./13 x 43 cm (medium); 5 x 17¾ in./13 x 45cm (large) onto card and cut out. Turn under the bottom edge of each cuff to the wrong side by ⅜ in. (1 cm) and pin in place. Straight stitch across by machine.

6 With right sides together, pin the cuffs around the boot warmers, aligning the raw edges at the top and the short sides of the cuffs with the inside seam of each boot warmer. Taking a ⅜-in. (1-cm) seam allowance, machine stitch around the top and down the short sides of the cuffs.

7 Turn to the wrong side so that the right sides of the cuff are facing out and the right sides of the boot warmers are facing in. Trim all seam allowances to ⅛ in. (3 mm) to reduce the bulk.

Making the bows

8 Cut out two strips of fleece measuring 2 x 27½ in. (5 x 70 cm). Fold in half lengthwise with right sides together, and pin along the long unfolded edge. Straight stitch along, leaving the sides open. Turn right side out.

9 Tie each fleece strip in a bow, tuck in the raw short ends of each bow by ⅜ in. (1 cm), and whipstitch (see page 113) the openings closed. Position a bow on the side of each boot warmer and hand stitch in place.

Why not try braiding (plaiting) strips of fleece together to make this fun neck-warmer design? There is only one short line of machine stitching involved, so it's a really quick-and-easy project.

simple snood

1 Cut out three strips of fleece, each measuring 5 x 27½ in. (12 x 70 cm). You may want to use two different shades or fleece, or simply stick to one shade. I've used one fuchsia pink strip and two orange strips.

2 Place the strips on top of each other, with the fuchsia pink strip sandwiched in the middle, and clip together onto a flat surface with a bulldog clip. Loosely braid (plait) the strips together.

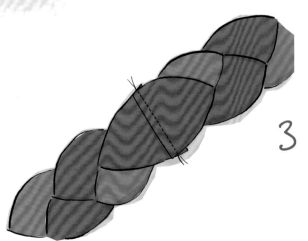

3 Remove the bulldog clip. Pin the bottom ends of the strips onto the top ends, so that you form a ring. Straight stitch across the join by machine.

You will need

40 x 60 in. (100 x
150 cm) fuchsia
pink fleece
Fuchsia pink polyester
sewing thread
Sewing machine
Basic sewing kit (see
page 111)

By simply folding and layering fleece fabric, you can create a super-stylish accessory; this beautiful and unique ruffled scarf will add chic to any outfit!

ruffled scarf

1 Cut out an 8 x 60-in. (20 x 150-cm) strip of fuchsia pink fleece for the main scarf shape and place it right side up on your work surface.

2 Cut out a 2¾ x 60-in. (7 x 150-cm) strip of fuchsia fleece. Place it right side up at one end of the scarf, centered on the width, and pin the strips together. Using both hands, make a fold ¾ in. (2 cm) deep and pin it in place.

3 Continue making folds and pinning them in place all the way along the scarf, adding a new 2¾ x 60-in. (7 x 150-cm) strip of fleece whenever you need to.

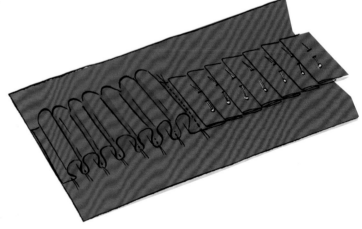

4 Straight stitch across each fabric fold by machine, removing the pins as you go.

You don't even need a pattern for this hat, so it's a great project for beginners. All you need are four rectangles of fabric and the ability to wield a pair of scissors and stitch in a (reasonably!) straight line.

fringe-topped hat

Skill level: 1

You will need
40 x 60 in. (100 x 150 cm) sage green fleece
Sage green polyester sewing thread
Sage green embroidery floss (thread)
Sewing machine
Basic sewing kit (see page 111)

To fit sizes
Small: hat diameter 22 in. (56 cm); height 7 in. (18 cm)
Medium: hat diameter 22¾ in. (58 cm); height 7½ in. (19 cm)
Large: hat diameter 23½ in. (60 cm); height 8 in. (20 cm)

1 Cut out four rectangles of sage green fleece of the appropriate size: small—12½ x 11¾ in. (32 x 30 cm); medium—13 x 12¼ in. (33 x 31 cm); large—13½ x 12½ in. (34 x 32 cm).

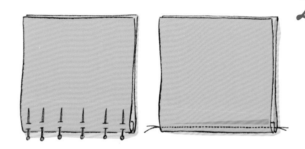

2 Place the fleece pieces together in pairs, wrong sides together. Tuck under both bottom edges by ¾ in. (2 cm) and pin to hold. Straight stitch across by machine, stitching ⅜ in. (1 cm) in from the bottom edge.

3 Place the paired-up fleece pieces right sides together. Fold under the bottom edge of each fleece section by 2 in. (5 cm). Pin at the sides and machine stitch, taking a ⅜-in. (1-cm) seam allowance. Trim the seam allowances to ⅛ in. (3 mm) to reduce the bulk. Turn right side out so that the seam isn't visible.

4 Cut slits 2¾ in. (7 cm) deep and ⅜ in. (1 cm) apart along the top open edge, cutting through all layers of fabric.

5 Gather up the cut strips in your hand and tie a small piece of embroidery floss (thread) tightly around the base.

Cheer yourself up over the winter months with these adorable super-snug fleece slippers—great for keeping feet toasty when you're lounging at home.

slipper socks

Skill level: 2

You will need
Templates on page 123
40 x 60 in. (100 x 150 cm) polka-dot yellow fleece fabric
10-in. (25-cm) square of felt, ⅛ in. (3 mm) thick
Yellow polyester sewing thread
20 in. (50 cm) elastic, ¼ in. (5 mm) wide
Sewing machine
Basic sewing kit (see page 111)

To fit ladies' shoe sizes
Small: US 5.5–6.5 (UK 3–4); 9¾ in. (25 cm) from heel to toe
Medium: US 7.5–8.5 (UK 5–6); 10¾ in. (27 cm) from heel to toe
Large: US 9.5–11.5 (UK 7–9); 11½ in. (29 cm) from heel to toe

1 Transfer the templates on page 123 onto card (see page 112) and cut out. Place the sole pattern on the wrong side of the polka-dot yellow fleece (on the grainline). Using a permanent marker pen, draw around it twice, flip the pattern over, and draw around it twice more. Cut out all four fleece sole pieces.

2 Place the inner pattern on thick felt, draw around it once, flip the pattern over, and draw around it again. Cut out both shapes.

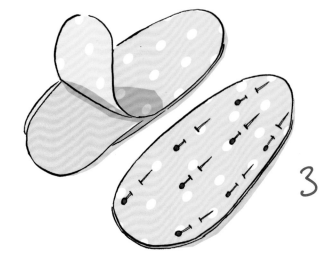

3 Pair up the left and right sole pieces with the right sides of the fleece facing inward and place the corresponding inner inside each one. Pin together at the center and sides.

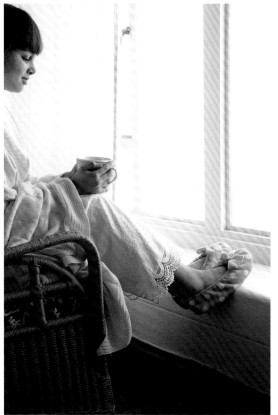

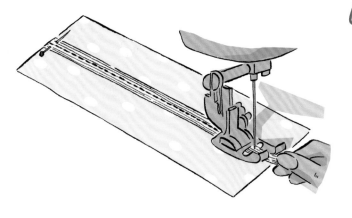

4 Depending on the size of slipper you are making, cut two strips of fleece fabric measuring 10¾ x 6 in. /28 x 15 cm (small), 11½ x 6 in./29 x 15 cm (medium), or 12¼ x 6 in./31 x 15 cm (large). With the fleece strip wrong side up, place a piece of elastic 8 in./20 cm (8¾ in./22cm; 9½ in./24 cm) in length, 2¾ in. (7 cm) down from the top edge. Pin at the start, stretch the elastic to fit across the whole width of the piece, and straight stitch across by machine.

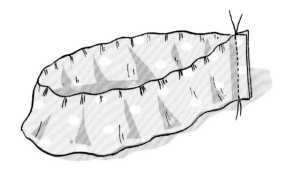

5 Fold each strip in half lengthwise, with wrong sides together, and machine stitch across the open side edges, taking a ⅜-in. (1-cm) seam allowance. Trim the seam allowances to ⅛ in. (3 mm) to reduce the bulk.

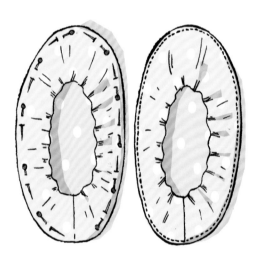

6 Pin an elasticated fleece strip around each slipper with the elasticated side facing inwards, making sure you align the stitched side seams with the center of the heels. Machine straight stitch around, taking a ⅜-in. (1-cm) seam allowance. Trim the seam allowances to ⅛ in. (3 mm) to reduce the bulk and turn each slipper sock right side out.

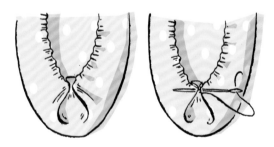

7 At the center front of each slipper, fold the surplus fabric down into a V-shape and whipstitch (see page 113) in place.

8 Cut two strips of fleece fabric measuring 10 x ⅝ in. (25 x 1.5 cm). Tie each one in a bow. Place one bow on the front of each slipper sock (top of fold) and whipstitch in place.

This large, stripy bag is not only great to use for those little trips to the supermarket, it's also big and sturdy enough to tote around as a day bag or take to the beach in the summer time.

tote bag

Skill level: 1

You will need
40 x 60 in. (100 x 150 cm) each of striped blue-and-white and plain lilac fleece
20-in. (50-cm) square each of fuchsia and sage green fleece
White polyester sewing thread
Sewing machine
Basic sewing kit (see page 111)

1 Cut two pieces measuring 16½ x 13¾ in. (42 x 35 cm) from striped blue-and-white fleece and two more from lilac fleece. Place a small plate or bowl on the wrong side of one short end of each piece and draw around it with a permanent marker pen. Cut along your drawn lines to create rounded ends.

2 Fold over the straight edges of each piece to the wrong side by 1¾ in. (4.5 cm) and pin to hold. Machine stitch along these edges ¼ in. (5 mm) in from the edge to form a hem.

3   Cut one 3 x 20-in. (8 x 50-cm) strip of fuchsia pink fleece, one 2½ x 16-in. (6 x 40-cm) strip of sage green fleece, and one 2 x 12-in. (4 x 30-cm) strip of lilac fleece. Work a line of running stitch (see page 112) along one short end of each fabric strip, pull tightly to pucker the fabric, and tie the thread in a knot.

4   Place the fabric strip on a flat surface and begin to twist it, using your right index finger to hold the center of the fabric rose and your left to twist the fabric. When you reach the end of the strip, tuck under the end of the strip and whipstitch it (see page 113) to the back of the rose.

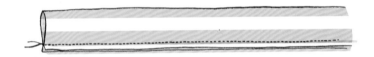
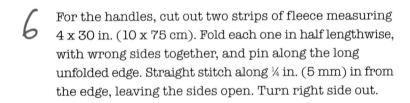

5   Using the photo as a guide, place the roses on the outer front of the bag and attach using fabric glue.

6   For the handles, cut out two strips of fleece measuring 4 x 30 in. (10 x 75 cm). Fold each one in half lengthwise, with wrong sides together, and pin along the long unfolded edge. Straight stitch along ¼ in. (5 mm) in from the edge, leaving the sides open. Turn right side out.

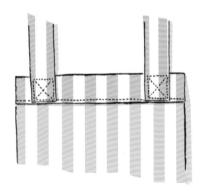

7   Place the outer front of the bag wrong side up on your work surface. Position the ends of one handle about 1½ in. (4 cm) in from each side edge, with the bottom of the handle level with the hem made in step 2. Pin in place. Box stitch the handle in place (see page 114). Repeat on the outer back of the bag.

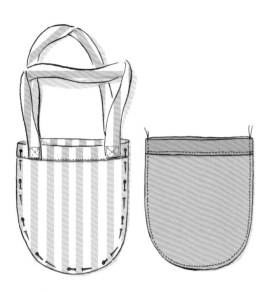

8 Place the outer bag front and bag pieces right sides together and pin to hold. Taking a ⅜-in. (1-cm) seam allowance, straight stitch around the bag. Turn right side out. Repeat with the lining pieces, but leave the lining wrong side out.

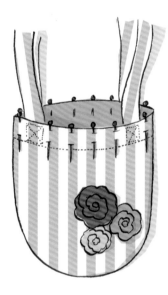

9 Push the lining inside the bag, carefully aligning the top edges and side seams, and make sure it lies flat. Pin around the top edge of the bag to hold the lining in place and then stitch around the top edge by machine.

Lightweight and comfortable, this tartan hat is the perfect accessory for cold winter days or chilly spring mornings. Because fleece doesn't wrinkle, you can stuff the hat into a pocket and it will still look great after you pull it out again! It is reversible, so you can wear it either tartan or lining side out.

tartan hat with ear flaps

Skill level: 2

You will need
Template on page 119
40 x 60 in. (100 x 150 cm) each of red tartan and purple fleece
Red polyester sewing thread
Purple embroidery floss (thread)

To fit sizes
Small: hat diameter 22 in. (56 cm); height 7 in. (18 cm)
Medium: hat diameter 22¾ in. (58 cm); height 7½ in. (19 cm)
Large: hat diameter 23½ in. (60 cm); height 8 in. (20 cm)

1 Transfer the template on page 119 onto card (see page 112) and cut out. Place the template on the wrong side of the red tartan fleece (on the grainline), close to the left-hand edge. Using a permanent marker pen, draw around it once. Flip the template over, butt the straight part of the hat section up against your previously drawn line, and draw around it again. Repeat for the lining, working on the wrong side of the purple fleece. Cut out all pieces; you will have one outer hat and one lining piece, both with four "peaks."

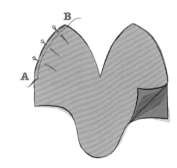

2 Fold the lining fleece piece in half at the center, with right sides together and the fold in the fabric on the left-hand side. Pin to hold. Machine stitch around the curve from A to B, working ⅛ in. (3 mm) in from the edge.

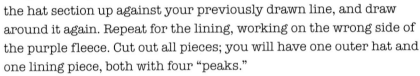

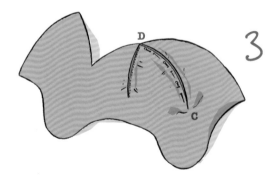

3 Open the piece out so that the previous seam is in the center, then fold over the two right-hand "peaks" so that their edges align. Pin to hold. Machine stitch around the curve from C to D, working ⅛ in. (3 mm) in from the edge.

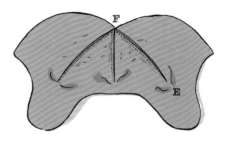

4 Repeat step 3 on the two left-hand "peaks," stitching the curve from E to F.

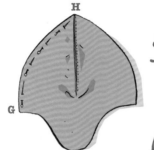

5 Finally, stitch the two open edges together between G and H to create the center back seam.

6 Repeat steps 2 through 5 with the tartan outer piece.

7 Place the lining inside the outer hat, with the wrong side of the lining against the right side of the tartan, carefully aligning the seams and unstitched edges. Pin together along the unstitched straight bottom edges. Machine stitch around the bottom edge, leaving a 3-in. (8-cm) opening at the bottom of one of the ear flaps for turning.

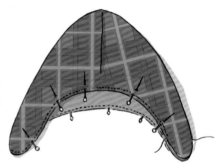

8 Pull the lining out from inside the hat. Turn both pieces through to the right side, through the opening at the back, push the lining into the outer fleece hat, and whipstitch (see page 113) the opening closed.

9 Cut a strip of purple fleece measuring 12 x 3 in. (30 x 8 cm). Cut slits 1⅜ in. (3.5 cm) long along both long edges, spacing them ⅜ in. (1 cm) apart. Roll up the fabric strip tightly, wrap a small piece of purple embroidery floss (thread) around the center, and tie tightly in a knot.

10 Position the pom-pom on the top of the hat and whipstitch it in place, using purple embroidery floss (thread).

Decorated with appliquéd orange mustaches and with a contrasting color of lining, these cozy mittens are both a fun and practical winter accessory!

mustache mittens

Skill level: 3

You will need
Templates on pages 126–127
40 x 60 in. (100 x 150 cm) each of beige and orange fleece
Sewing threads to match the fabrics
Sewing machine
Basic sewing kit (see page 111)
Darning foot

Cutting out the pieces

1 Transfer the templates for the outer and inner mittens on pages 126–127 onto card (see page 112) and cut out. Place template 1 (front) on the wrong side of the beige fleece (across the grainline). Using a permanent marker pen, draw around it twice. Repeat for the lining (inner) pieces, working on the wrong side of the orange fleece.

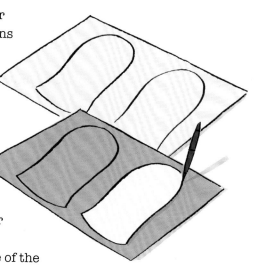

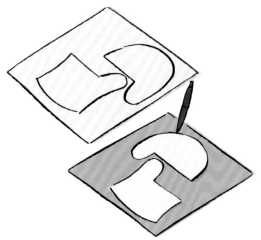

2 Place templates 2 and 3 (upper and lower back) on the wrong side of the beige fleece and draw around each one once for the right outer mittens. Repeat for the linings, working on the wrong side of the orange fleece.

3 Flip templates 2 and 3 over. Place them on the wrong side of the beige fleece and draw around each one once for the left outer mittens. Repeat for the linings, working on the wrong side of the orange fleece. Cut out all pieces.

Appliquéing the mustache

4 Transfer the mustache template on page 126 onto card (see page 112) and cut out. Place the template on the wrong side of the orange fleece, draw around it twice, and cut out the shapes.

5 Position the mustache shapes on the right and left front outer mittens and pin in the center to hold. Fit your sewing machine with a darning foot and free-motion stitch (see page 113) around each shape.

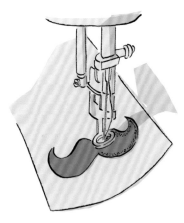

Assembling the mittens

6 Place the upper and lower back pieces right sides together in pairs; you will have a left and right outer back in beige and a left and right back lining in orange. Pin together. Taking a ⅜-in. (1-cm) seam allowance, machine stitch along the palms and around the thumbs. Trim the seam allowance to ⅛ in. (3 mm) to reduce the bulk.

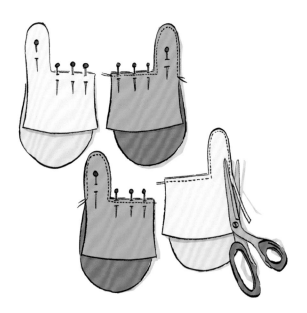

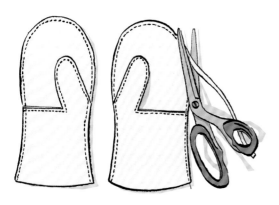

7   Pin the beige front and back outer pieces wrong sides together. Position the thumbs so they are sitting away from the edge and pin in place. Taking a ⅜-in. (1-cm) seam allowance, machine stitch all around, leaving the straight bottom edges open. Trim the seam allowances to ⅛ in. (3 mm) to reduce the bulk. Repeat with the orange lining pieces.

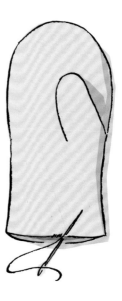

8   With the mittens wrong side out, match each outer mitten up with its lining, and pin together along the unstitched straight bottom edges. Machine stitch around the straight bottom edge, leaving a 2½-in. (6-cm) opening for turning at the back of the mittens. Turn right side out.

9   Turn right side out, push the lining into the outer mittens, and whipstitch (see page 113) the opening closed.

These sweet little mitts don't require a pattern so you'll have them made in no time! Keep them plain or jazz them up with pretty little embellished fleece flowers.

fingerless gloves

Skill level: 1

You will need
12 x 60 in. (30 x 150 cm) purple fleece
Purple polyester sewing thread
20 x 2 mm red glass beads
Sewing machine
Basic sewing kit (see page 111)

1 Cut out two strips of purple fleece fabric 11 in. (28 cm) long x 8½ in. (22 cm) wide, working along the grainline.

2 Fold each one in half lengthwise, wrong sides together. Tuck under both bottom open edges by ¾ in. (2 cm) and pin to hold. Straight stitch along by machine ⅜ in. (1 cm) in from the bottom edge, stitching through all layers. Trim the seam allowance to ⅛ in. (3 mm) to reduce the bulk.

3 Fold each strip in half widthwise and pin together along the side open edge. Taking a ⅜-in. (1-cm) seam allowance, machine stitch down this edge. Trim the seam allowance to ⅛ in. (3 mm) to reduce the bulk. Turn right side out.

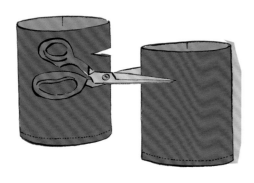

4 Place each piece so that the seam from the previous step is at the center back. Make a small snip big enough for your thumb to fit through at the right side of the left glove and the left side of the right glove, 2 in. (5 cm) down from the open top.

5 From the remaining fleece, cut two circles in each of the following sizes: 1½ in. (4 cm), 2 in. (5 cm), and 2½ in. (6 cm). Fold each circle in half, right sides together, and pin to hold. Make five small snips evenly around each half circle and cut out a tiny wedge shape at each point to create the petals.

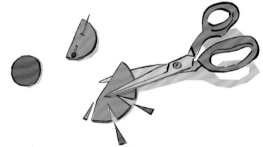

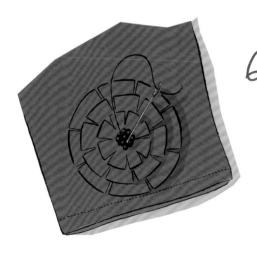

6 Layer three circles on top of each other in ascending order of size and position on the front of one glove, using the photo as a guide. Hand stitch ten beads to the center of the flower, making sure you only stitch through the front layer of the glove. Repeat with the remaining circles on the other glove.

This long stripy fleece scarf is a wonderful accessory for a man's coat or jacket and will look great with his winter wardrobe. The scarf is made up of fleece blocks that are sewn together and then doubled over. The double layer of fleece makes the scarf extra warm.

skinny stripy scarf

1 Cut out five 5½ x 7½-in. (14 x 19-cm) rectangles from each color of fleece.

Skill level: 1

You will need
10 x 30 in. (25 x 75 cm)
    each of beige,
    turquoise, and
    gray fleece
Beige polyester sewing
    thread
Sewing machine
Basic sewing kit (see
    page 111)

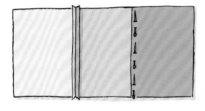

2 With right sides together, pin a gray rectangle to one long edge of a beige rectangle. Machine stitch, taking a ⅜-in. (1-cm) seam allowance. Attach a turquoise rectangle to the other side of the gray rectangle in the same way.

3 Repeat step 2 until you have used up all the pieces, maintaining the same color sequence throughout. Trim all the seam allowances to ⅛ in. (3 mm) to reduce the bulk.

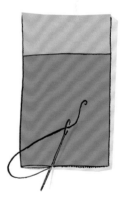

4 Fold the scarf in half lengthwise, right sides together. Pin along the long open edge and one side edge and machine stitch, taking a ⅜-in. (1-cm) seam allowance. Trim the seam allowance to ⅛ in. (3 mm) to reduce the bulk. Turn right side out.

5 Tuck in the open side edge by ⅜ in. (1 cm) and whipstitch (see page 113) the opening closed.

Using just one strip of fleece, you can create this pretty little headband in no time at all. The bow detail makes the design really special and the doubled-up fleece makes it extra cozy to wear!

bow headband

Skill level: 1

You will need

24-in. (60-cm) square
  of red fleece
Dark pink polyester
  sewing thread
Sewing machine
Basic sewing kit (see
  page 111)

To fit sizes

Small: headband
  circumference
  22 in. (56 cm)
Medium: headband
  circumference
  22¾ in. (58 cm)
Large: headband
  circumference
  23½ in. (60 cm)

Making the headband

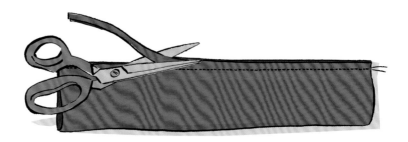

1 Cut a strip of red fleece on the grainline to the size required: 22¾ x 8 in./58 x 20 cm (small); 23½ x 8¾ in./60 x 22 cm (medium); or 24½ x 9½ in./62 x 24 cm (large). Fold the strip in half lengthwise, with right sides together, and pin to hold. Machine stitch across the long unfolded edge ⅜ in. (1 cm) from the edge. Trim the seam allowance to ⅛ in. (3 mm) to reduce the bulk.

2 Turn right side out and fold in half widthwise, making sure that the inside seam is positioned at the center of the fabric. Pin the headband together along the open side seam and machine stitch ⅜ in. (1 cm) from the edge. Trim the seam allowance to ⅛ in. (3 mm) to reduce the bulk.

3 Turn the side seam to the inside of the headband, so that it is not visible, and machine stitch around the top and bottom edges, ⅜ in. (1 cm) from the edge.

Making the bow

4   For the bow, cut two 3 x 5½-in.
(8 x 14-cm) strips of fleece.
Pin them wrong sides together
and work a machine zig-zag
stitch all around the edges.
Work a line of running stitch
(see page 112) by hand
through the center, then pull
both ends of the thread to
gather the fabric.

5   Cut a ¾ x 1¼-in. (2 x 3-cm)
strip of fleece, wrap it around
the center of the bow, and pull
it tightly around to the back
of the bow. Whipstitch (see
page 113) the top and bottom
edges of the strip to the back
layer of the fleece bow.

6   Pin the fleece bow to the center left side of the
headband and whipstitch it in place along the top and
bottom edges of the center strip, making sure you only
stitch through the top layer of the headband.

Chapter 2

For kids

Fleece is an amazing fabric to use for making children's accessories, as it's both ultra soft and durable—and because it's a synthetic fabric, it washes really easily and also dries quickly. I've tried to make each design in this chapter as colorful as possible, so you will find them super fun to make—my favorites are the cheeky froggy slippers and the squeezable owl cushion! So why not pick from the many designs and stitch up unforgettable gifts for your little ones?

Make these adorable mitts and help keep kids' hands warm as they play outdoors. Adding the lining makes them twice as cozy.

dinosaur mittens

Cutting out the mitten pieces

1 Transfer the templates on page 117 onto card (see page 112) and cut out. Place template 1 (front) on the wrong side of the orange fleece, making sure that you place it across the grainline. Using a permanent marker pen, draw around it twice for the front outer mittens. Repeat for the lining pieces, working on the wrong side of the red fleece.

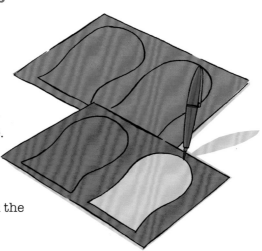

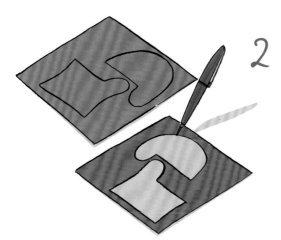

2 Place templates 2 and 3 (the thumb and lower back) on the wrong side of the orange fleece and draw around each one once for the right outer mittens. Repeat for the linings, working on the wrong side of the red fleece.

3 Flip templates 2 and 3 over. Place them on the wrong side of the orange fleece and draw around each one once for the left outer mittens. Repeat for the linings, working on the wrong side of the red fleece. Cut out all the pieces.

Skill level: 3

You will need
Templates on page 117
20-in. (50-cm) square each of red and orange fleece
6-in. (15-cm) square each of sage green and dark gray felt
Orange polyester sewing thread
8 (10:11) in./20 (24: 28) cm elastic, ¼ in. (5 mm) wide
Four googly eyes, approx. ¼ in. (7 mm) in diameter
Sewing machine
Basic sewing kit (see page 111)

To fit sizes
Small (age 3–5 years): 3⅛ in. (8 cm) wide x 6¼ in. (16 cm) long
Medium (age 6–8 years): 3½ in. (9 cm) wide x 7 in. (18 cm) long
Large (age 9–12 years): 4 in. (10 cm) wide x 8 in. (20 cm) long

Decorating the mittens

4 Trace the spot templates on page 117 onto card and cut out. Place the medium template on gray felt and the large template on green felt, draw around each one six times, and cut out. Place the medium template on green felt and the small template on gray felt, draw around each one four times, and cut out.

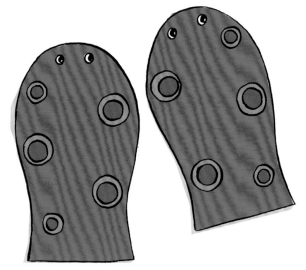

5 Place the small gray spots on the medium green spots, and the medium gray spots on the large green spots, and stick in place with fabric glue. Using the photo as a guide, place two medium and three large spots on each mitten front and glue in place. Position two googly eyes on each mitten front and glue in place.

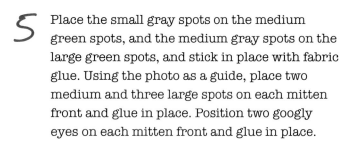

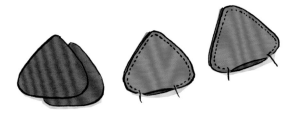

6 Trace the spike template on page 117 onto card and cut out. Place the template on the wrong side of the red fleece, draw around it 12 times, and cut out all the shapes. With right sides together, place the spikes together in pairs and machine stitch around two sides, leaving the bottom edge of each pair unstitched for turning. Turn the spikes right side out.

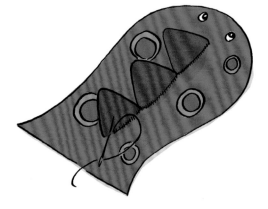

7 Turn under the raw edges. Pin three spikes along the center of each mitten front and whipstitch (see page 113) in place.

Assembling the mittens

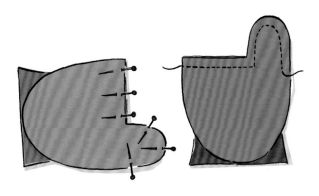

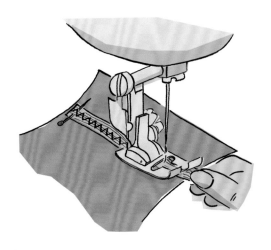

8 Place the thumb and lower back pieces right sides together in pairs, matching the thumbs; you will have a left and a right outer back in orange and a left and a right lining back in red. Pin together. Taking a ⅜-in. (1-cm) seam allowance, machine stitch along the palms and around the thumbs. Trim the seam allowance to ⅛ in. (3 mm) to reduce the bulk.

9 Pin a 2- (2½-: 2¾-) in./5- (6-: 7-) cm length of elastic to the wrong side of the orange front and back pieces of each mitten, 2 in. (5 cm) from the bottom edge. Using a wide zig-zag stitch, machine stitch across, stretching the elastic to fit across the whole width of the piece as you go.

10 Pin the orange front and back pieces right sides together. Taking a ⅜-in. (1-cm) seam allowance, machine stitch all around, leaving the straight bottom edges open. Trim the seam allowances to ⅛ in. (3 mm) to reduce the bulk. Repeat with the red lining pieces.

11 With the mittens wrong side out, match each outer mitten up with its corresponding lining, and pin together along the unstitched straight bottom edge. Machine stitch around the straight edge, leaving a 1½-in. (4-cm) opening for turning.

12 Turn right side out, push the lining into the outer mittens, and whipstitch the opening closed.

Your child will love this super-comfy beanie with its big ear flaps — it's just the thing for a chilly winter's day. The wooly pom-pom adds a cute finishing touch.

pom-pom beanie

Skill level: 3

You will need

Templates on page 121
1 yd (1 m) each of dark turquoise and lime-green fleece, 60 in. (150 cm) wide
Turquoise polyester sewing thread
Turquoise embroidery floss (thread)
Lime-green and turquoise yarn for pom-pom
Sewing machine
Basic sewing kit (see page 111)

To fit sizes

Small (age 3–5 years): hat circumference 19¼ in. (49 cm)
Medium (age 6–8 years): hat circumference 20 in. (51 cm)
Large (age 9–12 years): hat circumference 21¼ in. (54 cm)

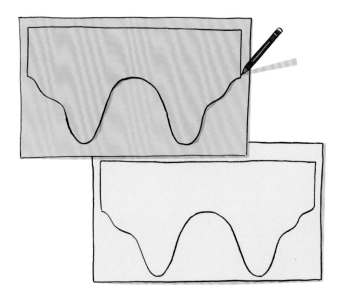

1 Transfer the templates on page 121 onto card (see page 112) and cut out. Place template 1 (the band and ear flaps) on the wrong side of the dark turquoise fleece, making sure you place it across the grainline. Using a permanent marker pen, draw around it once. Repeat for the lining, working on the wrong side of the lime-green fleece.

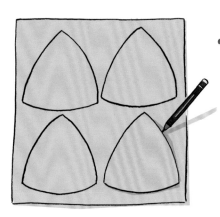

2 Place template 2 (the hat top section) on the wrong side of the dark turquoise fleece and draw around it four times.

3 Cut out all the pieces for the hat top, band, and ear flaps.

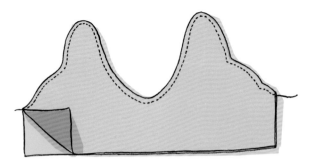

4 Pin the dark turquoise and lime-green band and ear flap pieces right sides together. Taking a ⅜-in. (1-cm) seam allowance, machine stitch around the ear flap sections, leaving the top and sides of the band open. Turn right side out.

5 With right sides together, pin the hat top sections together along their side edges. Taking a ⅜-in. (1-cm) seam allowance, machine stitch down each section.

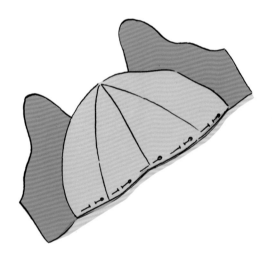

6 With right sides together, making sure that the edges align, pin the bottom edge of the hat top pieces to the top of the open band. Taking a ⅜-in. (1-cm) seam allowance, machine stitch the pieces together.

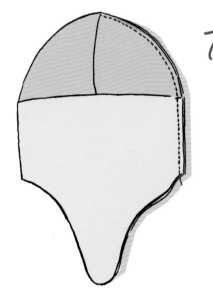

7 Fold the beanie in half, with the green lining facing outward, and pin the hat together along the back open seam. Machine stitch the back seam. Trim all seam allowances to ⅛ in. (3 mm) to reduce the bulk. Turn right side out.

8 Cut two 3-in. (8-cm) circles of card, with a small circular hole in the center of each. Place them together and wind the turquoise and lime-green yarn through the holes in the middle of the two pieces of card and around the outer circles. Repeat until you have the required amount of yarn wound around both circles. Then, using a pair of scissors, cut the yarn between both circles of card so that the loops are cut open. Open out the two circles slightly and tie a piece of strong embroidery floss (thread) tightly around the middle of the pom-pom and remove the card.

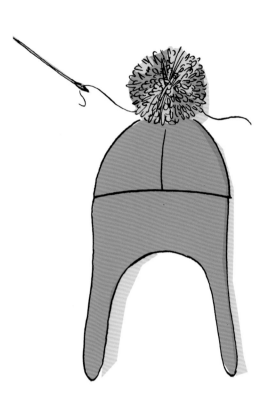

9 Place the pom-pom on the top of the hat and whipstitch it in place (see page 113), using the remainder of the embroidery floss (thread).

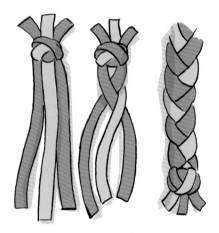

10 Cut six 1½ x 16-in. (4 x 40-cm) lengths of fleece—two dark turquoise and four lime-green. Separate into two groups of three strands each (one turquoise and two lime-green) and braid (plait) them tightly. Secure the end with a knot.

11 Place a braided (plaited) cord on the bottom of each ear flap and whipstitch in place, using matching sewing thread.

These cheeky little hug-me frog slippers, which can be worn by boys and girls alike, are a joy to make and will keep those little toes all snug before bedtime.

frog slippers

Skill level: 3

You will need

Templates on page 122
20-in. (50-cm) square each of emerald green and fluffy cream fleece
12-in. (30-cm) square of thick felt, ¼ in. (5 mm) thick
One 6-in. (15-cm) square each of white, brown, and red felt
Emerald green polyester sewing thread
Black embroidery floss (thread)
Sewing machine
Basic sewing kit (see page 111)

To fit children's shoe sizes

Small (age 3–5 years): US 9.5–10.5 (UK 9–10); 6¾ in. (17 cm) from heel to toe
Medium (age 6–8 years): US 11.5–13.5 (UK 11–12); 8 in. (20 cm) from heel to toe
Large (age 9–12 years): US 1.5–2.5 (UK 1–2); 8¾ in. (22 cm) from heel to toe

Making the slippers

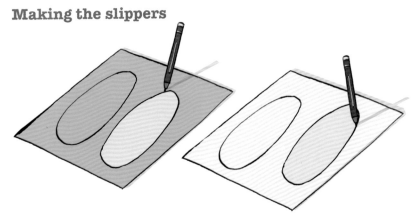

1 Transfer the templates on page 122 onto card (see page 112) and cut out. Place template 1 (sole) on the wrong side of the emerald green fleece and draw around it twice, using a permanent marker pen. Repeat on the wrong side of the fluffy cream fleece.

2 Place template 2 (slipper front) on the wrong side of the emerald green fleece and draw around it twice, using the permanent marker pen. Cut out all pieces.

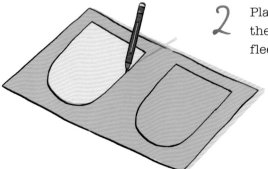

3 Place the bottom and top sole pieces wrong sides together and pin to hold. Taking a ⅜-in. (1-cm) seam allowance, work a small machine zig-zag stitch around the backs of the soles, leaving the front section open.

4 Place template 3 (slipper inner) on the thick felt fabric, draw around it twice, and cut out. Slot the felt inners inside the openings in the sole pieces.

5 Fold over the straight edges of the slipper fronts to the wrong side by 1¼ in. (3 cm) and pin in place. Straight stitch along this edge to form a hem.

Decorating the slippers

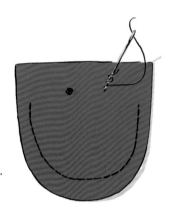

6 Using the photo as a guide, work a line of black backstitch (see page 112) by hand on each front piece to form the mouth and two French knots for the nostrils (see page 113).

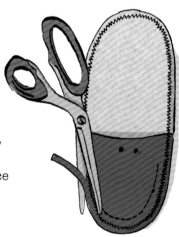

7 Place each front piece right side up on top of the cream side of the sole (over the opening left in step 4), pin together, and work a small machine zig-zag stitch around the curved front edge. Trim the seam allowance to ⅛ in. (3 mm) to reduce the bulk.

8 Using the templates on page 122, trace the frog eyes, hands, and tongue onto card and cut out. Place the outer eye template on green fleece, draw around it eight times, and cut out. Place the medium inner eye template on white felt and the small inner eye template on brown felt. Draw around each one four times and cut out.

9 Place each brown eye on a white eye piece, and pin to hold. Straight stitch all around by hand or by machine. Place each assembled inner eye piece on a green outer eye and straight stitch all around by hand or by machine.

10 With right sides together, place a plain green outer eye on top of each assembled eye from the previous step. Straight stitch around the edges by machine, leaving a small opening at the bottom for turning.

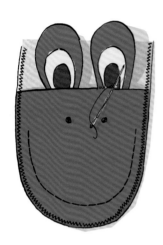

11 Turn each eye right side out. Using the photo as a guide, place the eyes on the tops of the front pieces, and whipstitch (see page 113) in place.

12 Place the hand template on green fleece, draw around it eight times, and cut out.

13 Place the hand pieces together in pairs, with right sides together, and straight stitch around the edges, leaving a small opening at the bottom for turning. Turn right side out and whipstitch the opening closed. Using the photo as a guide, place the hands on the sides of the slipper fronts and whipstitch in place.

14 Place the tongue template on red felt, draw around it twice, and cut out the shapes. Place under the embroidered mouth and whipstitch in place.

This charming snake scarf features lots of fun zig-zag decorations. They are so straightforward to appliqué and stitch, yet the effect is striking. Sweet and playful, this scarf is ideal for brightening up your little one's wardrobe.

snake scarf

Skill level: 1

You will need
Templates on page 121
2 yd (2 m) beige fleece, 60 in. (150 cm) wide
30 in. (75 cm) each of green and red ric-rac braid
9-in. (23-cm) square each of green, light blue, yellow, purple, and red felt
Sewing threads to match the fleece and felts
2 x 5 mm black beads
Sewing machine
Basic sewing kit (see page 111)
Adhesive tape

1   Transfer the head and tail templates on page 121 onto card and cut out. Cut out a rectangle 51 x 7 in. (130 x 18 cm) for the body template. Tape them together to form the snake shape. Pin to the wrong side of the beige fleece. Draw around it with a permanent marker pen and cut out the front of the snake. Repeat for the back of the snake.

2   Cut four 7-in. (18-cm) lengths each of red and green ric-rac braid. Using the photo as a guide, pin them in pairs to the right side of the snake front. Machine stitch along the center of each length.

3   Trace the zig-zag template on page 121 onto card, draw around it, and cut it out. Place on the felt fabrics, draw around it with a marker pen, and cut out one light blue, one yellow, and two purple zig-zags.

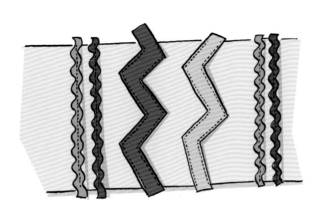

4 Using the photo as a guide, pin the felt zig-zags to the right side of the snake front. Machine stitch around each one, working ¹⁄₁₆ in. (2 mm) in from the edges.

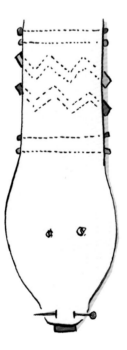

5 Hand stitch two black beads to the center of the front head piece for eyes.

6 Trace the tongue template on page 121 onto card and cut out. Place it on red felt, draw around it, and cut out the shape.

7 Place the front and back of the snake right sides together. Tuck the tongue under the top of the snake head and pin in place.

8 Taking a ⅜-in. (1-cm) seam allowance, straight stitch around the snake, leaving a 3-in. (8-cm) gap at the bottom of the body for turning.

9 Trim the seam allowance to ⅛ in. (3 mm) to reduce the bulk, turn right side out, and whipstitch (see page 113) the opening closed.

This simple-to-make cozy with fleecy pom-poms is great for snuggling up to on a cold winter night. Have fun adding some extra embellishment with a bit of freehand embroidered detail.

hot-water bottle cozy

Skill level: 2

You will need
Templates on page 115
1 yd (1 m) mustard-
   yellow fleece, 60 in.
   (150 cm) wide
10 in. (25 cm) pink
   fleece, 28 in. (70 cm)
   wide
6-in. (15-cm) square
   each of mauve and
   white felt
10 in. (25 cm) elastic,
   ¼ in. (5 mm) wide
Three small purple
   buttons, ⅝ in. (1.5 cm)
   in diameter
Mustard-yellow
   polyester sewing
   thread
Pink machine
   embroidery floss
   (thread)
Sewing machine
Basic sewing kit (see
   page 111)
Embroidery hoop
Darning foot

**To fit a small
hot-water bottle**

Making the cozy

1. From mustard-yellow fleece, cut two rectangles measuring 11 x 7½ in. (28 x 19 cm) for the front and back of the cozy.

2. Fold the top edge of the front and back pieces over to the wrong side by ⅜ in. (1 cm) and straight stitch across.

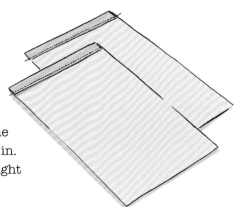

3. Cut the elastic in half. Place one piece on the wrong side of each fleece piece 4 in. (10 cm) from the top. Pin the elastic at one side edge to hold it in place. Using a wide zig-zag stitch, machine stitch across, stretching the elastic to fit across the whole width of the piece as you go.

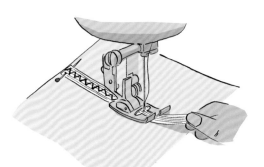

Decorating the cozy

4. Trace the snowflake templates on page 115 onto card (see page 112) and cut out. Place the small template on white felt, the medium template on the wrong side of the pink fleece, and the large template on mauve felt. Draw around the templates and cut out all the shapes.

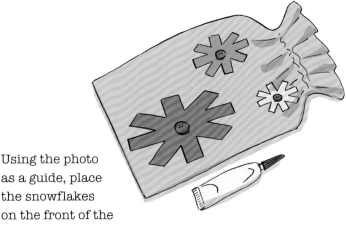

5. Using the photo as a guide, place the snowflakes on the front of the hot-water bottle cozy and attach using fabric glue. Glue a button to the center of each snowflake.

6 Using the photo as a guide, draw snowflake designs on the fleece with a permanent marker pen. Replace your regular sewing-machine foot with a darning foot and thread the machine with pink machine embroidery floss (thread). Set the stitch tension to 0 and drop the feed dogs. Place the fabric in an embroidery hoop and free-motion stitch (see page 113) over your snowflake drawings.

7 Re-thread the machine with mustard-yellow sewing thread. Place the front and back pieces of the cozy right sides together and pin around the side edges and base. Machine stitch along the sides and base, taking a ⅜-in. (1-cm) seam allowance. Trim the seam allowance to ⅛ in. (3 mm) to reduce the bulk. Turn right side out.

8 Cut three 12 x 1¼-in. (30 x 3-cm) strips of mustard-yellow fleece. Knot all three together at one end, braid (plait) them together, and knot the other end to secure.

9 Cut two strips of pink fleece measuring 12 x 3 in. (30 x 8 cm). Make 1¼-in. (3-cm) cuts, ⅜ in. (1 cm) apart, along both long edges. Roll up the fabric strips tightly, wrap a small piece of embroidery floss (thread) around the middle of each one, and tie tightly in a knot.

10 Hand stitch one pom-pom to one end of each braided (plaited) cord and whipstitch (see page 113) the cords to one side of the cozy.

Lovely and pretty, this lightweight and soft necklace is really comfortable to wear and easy for a little girl to pull on and off. It could even be worn as a bracelet or turned into a flower brooch!

flower necklace

Skill level: 1

You will need
Template on page 116
10-in. (25-cm) square
    of lilac fleece
Six purple and nine pink
    5 mm plastic beads
32 in. (80 cm) light blue
    organza ribbon, ⅝ in.
    (1.5 cm) wide
Lilac polyester sewing
    thread
Clamshell ribbon ends
    and a hook-and-eye
    fastener (optional)
Sewing machine
Basic sewing kit (see
    page 111)

Making the flower

1   Transfer the petal template on page 116 onto card (see page 112) and cut out. Place the template on the wrong side of the lilac fleece, draw around it ten times with a permanent marker pen, and cut out all the petals.

2   Pin the petals together in pairs, with right sides together. Machine stitch around the curved edge, leaving the bottom straight edge open. Turn right side out.

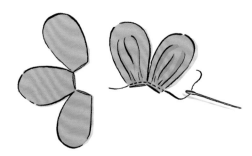

3   Group the petals together in a cluster of five. Using two strands of thread, work a line of running stitch (see page 112) through all open bottom edges to join the petals together. Stitch three pink beads to the center of the flower.

Assembling the necklace

4   Thread the ribbon through a large-eyed embroidery needle. Pull the needle through six of the remaining plastic beads, alternating the colors. Leave a 2¾-in. (7-cm) gap in the middle for the fleece flower, then thread on the remaining beads, reversing the color sequence of the first group. Tie a double knot where the first and last beads are positioned to ensure that they don't fall off the ribbon.

5   Place the flower in between the two groups of beads and whipstitch (see page 113) the left and right petals and the center of the flower to the ribbon. To secure further, cut a ¾ x 1¼-in. (2 x 3-cm) square of fleece and glue it over the center back of the flower, on top of the ribbon.

6   To wear the necklace, either tie the ends of the ribbon together or attach clamshell ribbon ends and a hook-and-eye fastener.

This cheeky monkey hat is a really fun project to make. The lining makes it extra warm, so your little one can monkey around outdoors to their heart's content!

cheeky monkey hat

Skill level: 2

You will need
Templates on page 120
1 yd (1m) each of lilac
   and white fleece, 50 in.
   (150 cm) wide
12-in. (30-cm) square of
   red felt
4-in. (10-cm) square of
   pink polka-dot fabric
Small sheet of fusible
   bonding web
Sewing threads to match
   the fleece fabrics
Red embroidery floss
   (thread)
Two small black buttons,
   ⅜ in. (8 mm) in diameter
Sewing machine
Basic sewing kit (see page
   111)
Darning foot

To fit sizes
Small (age 3–5 years):
   hat circumference
   19¼ in. (50 cm)
Medium (age 6–8 years):
   hat circumference
   20 in. (53 cm)
Large (age 9–12 years):
   hat circumference
   21¼ in. (55 cm)

Making the hat pieces

1 Transfer the templates for the outer hat and the lining on page 120 onto card (see page 112) and cut out. Place the template on the wrong side of the lilac fleece (on the grainline). Using a permanent marker pen, draw around it twice. Repeat for the lining, working on the wrong side of the white fleece. Cut out all four pieces.

2 Fold each fleece piece in half, with right sides together, so that the curved edges align. Pin to hold, then machine stitch along the curved inner edge of each piece, working close to the edge. Cut along the straight part of the edge opposite the one you stitched, so that you get a semicircle shape.

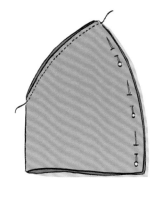

3 Open out one lilac piece; this will be the front of the hat—the monkey's face. Transfer the templates for the outer and inner mouths on page 120 onto card and cut out. Place the outer mouth template on the wrong side of the remaining white fleece, draw around it once with a permanent marker pen, and cut out the shape. Place the inner mouth template on red felt, draw around it once, and cut out the shape.

4 Using the photo as a guide, place the outer mouth on the front fleece piece, pin in place, and machine stitch all around. Place the inner mouth on top of the outer mouth and attach using fabric glue.

5 Trace the template for the heart on page 120 onto the paper side of the fusible bonding web and cut out roughly (see page 114). Press in place on the wrong side of the pink polka-dot fabric, cut along the drawn lines, and peel off the backing paper.

6 Place the heart on the front fleece piece, cover with a piece of parchment paper or a clean dish towel, and press for a few seconds to fuse the fabrics together, making sure your iron is at a low setting.

7 Fit a darning foot to your machine and free-motion stitch (see page 113) around the heart.

Decorating the hat

8 Place two buttons on the front fleece piece for the eyes and hand stitch in place. Using the photo as a guide and red embroidery floss (thread), work two cross stitches on the front fleece piece for the nose.

9 Trace the template for the ears on page 120 onto card and cut out. Place on the wrong side of the lilac fleece, draw around four times, and cut out all the shapes.

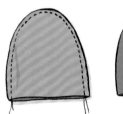

10 Pin the ear pieces together in pairs, with right sides together, and machine stitch around the curved edges. Turn right side out.

Assembling the hat

11 Place the outer hat pieces right sides together. Position one ear inside each side edge of the hat, and pin in place. Taking a ⅜-in. (1-cm) seam allowance, machine stitch around the sides, leaving the straight bottom edge open. Repeat with the lining pieces. Trim all seam allowances to ⅛ in. (3 mm) to reduce the bulk.

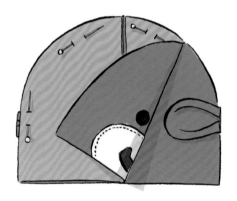

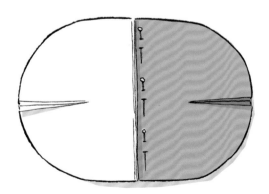

12 Pin the outer and lining hats together, still wrong side out, along the unstitched straight edges, with the outer hat overlapping the lining. Straight stitch all around the unstitched straight edges, leaving a 2½-in. (6-cm) opening for turning.

13   Turn right side out, push the lining into the outer hat, and whipstitch (see page 113) the opening closed.

14   For the hair, cut a 10 x 5-in. (25 x 13-cm) strip of red felt. Make cuts 2¼ in. (5.5 cm) deep, ⅜ in. (1 cm) apart along both long edges. Roll up the fabric strip tightly, wrap a small piece of embroidery floss (thread) around the middle, and tie tightly in a knot.

15   Hand stitch the hair to the top of the hat, using two strands of red embroidery floss (thread).

This adorable cat bag has been made using super-cute pink polka-dot fleece fabric. The fleece lining gives it extra strength, making it the perfect accessory for any little girl to carry around her bits and bobs.

cat bag

Skill level: 2

You will need
Templates on page 116
1 yd (1 m) pink-and-
    white polka-dot fleece,
    60 in. (150 cm) wide
1 yd (1 m) white fleece,
    60 in. (150 cm) wide,
    for the lining
6-in. (15-cm) square
    of pale pink felt
2 black buttons, ⅜ in.
    (1 cm) in diameter
Blue tapestry yarn
Purple embroidery
    floss (thread)
Pink polyester sewing
    thread
Sewing machine
Basic sewing kit (see
    page 111)

Preparing the pieces

1 From pink-and-white polka-dot fleece, cut two rectangles measuring 4⅜ x 4¾ in. (11 x 12 cm). Place a small plate or bowl on the wrong side of each short end and draw around it with a permanent marker pen. Cut along your drawn lines to create rounded ends. Repeat for the lining, working on the wrong side of the white fleece.

2 Fold the straight edge of each polka-dot piece over to the wrong side by 1¾ in. (4.5 cm) and pin in place. Machine straight stitch along these edges to form a hem.

Decorating the bag

3 Place one outer fleece piece (the bag front) right side up on your work surface. Using a permanent marker pen and referring to the photo as a guide, mark out the cat nose shape. Using blue tapestry yarn, work satin stitch (see page 113) across the nose. Using purple embroidery floss (thread), work backstitch (see page 112) across the mouth and three 2-in. (5-cm) long lines of backstitch on either side of the nose for the whiskers.

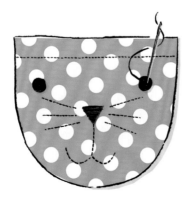

4 Using the photo as a guide, place two black buttons above the cat's nose in line with the outer edge of the whiskers and hand stitch in place with purple embroidery floss (thread) to form the eyes.

5 Transfer the templates on page 120 onto card (see page 112) and cut out. Place the outer ear template on the right side of the pink polka-dot fleece, draw around it four times, and cut out. Place the inner ear template on pale pink felt, draw around it twice, and cut out.

6 Place the felt inner ears on the wrong side of two of the fleece outer ear pieces, aligning them at the base. Pin in place and machine stitch all around. These are the fronts of the ears.

7 With right sides together, pin each front ear to one of the remaining fleece ear pieces and machine stitch all around the edges.

Assembling the bag

8 Cut a strip of fleece measuring 8 x 2¾ in. (20 x 7 cm). Fold it in half lengthwise, with wrong sides together, and pin to hold. Machine stitch around, leaving a small gap at the bottom for turning.

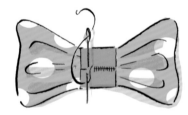

9 Turn through, so that the strip is wrong side out, and whipstitch (see page 113) the opening closed. Work a line of running stitch (see page 112) by hand through the center, then pull the threads to gather the fabric. Cut a ⅝ x 1½-in. (1.5 x 4-cm) strip of pale pink felt, wrap it around the center of the bow, and pull it tightly around to the back of the bow. Whipstitch the top and bottom edges of the strip to the back layer of the fleece bow.

10 For the handles, cut out two strips of fleece measuring 4 x 24 in. (10 x 60 cm). Fold each one in half lengthwise with wrong sides together, and pin along the long unfolded edge. Machine stitch along this edge, leaving the sides open. Turn right side out.

11 Position the ends of one handle on the wrong side of the front of the bag, about 1½ in. (4 cm) in from each side edge and within the hem that you made in step 2, and pin in place. Machine stitch a box shape on each handle end (see page 114). Repeat for the back of the bag.

| 2 Pin the front and back of the polka-dot outer bag right sides together. Taking a ⅜-in. (1-cm) seam allowance, machine stitch around the curved sides and base of the bag. Turn right side out.

| 3 Repeat step 12 with the white fleece lining pieces, but do not turn the lining right side out. Fold the top of the bag lining over to the wrong side by 1¾ in. (4.5 cm).

| 4 Insert the lining into the bag, with wrong sides together, and make sure it lies flat. Pin around the top edge of the bag to hold the lining in place and then machine stitch around the top edge.

| 5 Using the photo as a guide, pin the ears and bow onto the bag and whipstitch in place.

Your little one will love cozying up to this huggable owl toy, which can double up as a cute pillow for their bedroom. The free-motion stitch technique means this project is super quick to make.

owl toy

Skill level: 2

You will need
Templates on page 120
1 yd (1 m) gray fleece,
    60 in. (150 cm) wide
12-in. (30-cm) square
    each of white, red
    polka-dot, and
    mustard yellow fleece
20-in. (50-cm) square of
    turquoise fleece
Lilac polyester sewing
    thread
Turquoise polyester
    sewing thread
Toy filling
Sewing machine
Basic sewing kit (see
    page 111)
Darning foot

Stitching the owl

1 Transfer the templates on page 120 onto card (see page 112) and cut out. Place the body template on the wrong side of the gray fleece (on the grainline). Using a permanent marker pen, draw around it twice. Cut out both pieces.

2 Using the templates for the eyes, heart, beak, wings, and feathers, draw around the templates on the wrong side of the relevant color of fleece: cut out two eyes from white fleece, one heart from red polka-dot fleece, two wings from turquoise fleece, and one beak from mustard yellow fleece.

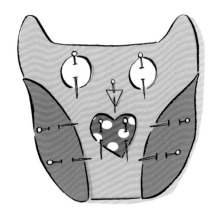

3 Place the owl body front right side up on your work surface. Using the photo as a guide, pin the eyes, heart shape, wings, and beak in place.

4 Fit a darning foot on your sewing machine and thread the machine with lilac polyester thread. Using the photo as a guide, free-motion stitch (see page 113) the eye design and around the edge of the beak and heart.

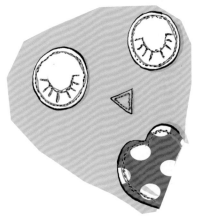

5 Change to turquoise thread. Using the photo as a guide, free-motion stitch around each wing and stitch a small heart design in the center of the large polka-dot heart.

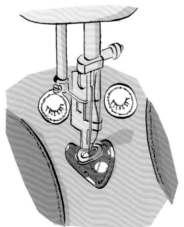

6 Place the feather template on the wrong side of the mustard yellow fleece, draw around it twice, and cut out the shapes. Using the photo as a guide, position the feathers in the center of the wings and attach using fabric glue.

Assembling the toy

7 Change back to a regular sewing machine foot. Pin the front and back of the owl right sides together. Taking a ⅜-in. (1-cm) seam allowance, machine stitch around the owl, leaving a 3-in. (8-cm) gap at the bottom for turning.

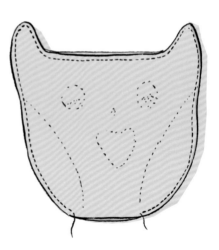

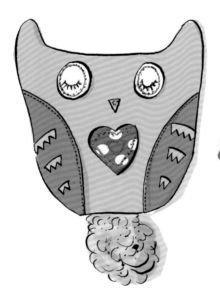

8 Trim the seam allowance to ⅛ in. (3 mm) to reduce the bulk, and turn the owl right side out. Stuff with toy filling and whipstitch (see page 113) the opening closed.

With hardly any sewing involved, this is a really simple project to make for a little girl or boy. Kids will love how easy it is to wrap it around their neck, with no fleecy fringes to get all tangled up in!

woven scarf

Skill level: 1

You will need

20 in. (50 cm) each of light blue and white fleece, 60 in. (150 cm) wide

10 in. (25 cm) lilac fleece, 60 in. (150 cm) wide

White embroidery floss (thread)

Basic sewing kit (see page 111)

1 Cut out a 6¼ x 55-in. (16 x 140-cm) strip of light blue fleece.

2 Fold the long edges in to the center and pin in place. Working along the folded edges, cut slits ¾ in. (2 cm) long every 2¾ in. (7 cm).

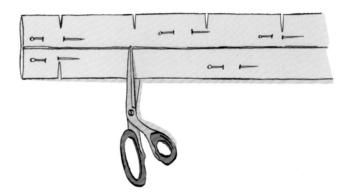

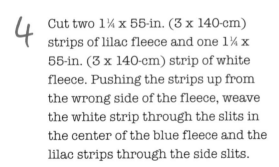

3 Open out and fold the strip in half lengthwise, wrong sides together. Pin to hold. Cut a slit in the folded edge ¾ in. (2 cm) long and 1⅜ in. (3.5 cm) from the side edge. Continue to cut slits every 2¾ in. (7 cm).

4 Cut two 1¼ x 55-in. (3 x 140-cm) strips of lilac fleece and one 1¼ x 55-in. (3 x 140-cm) strip of white fleece. Pushing the strips up from the wrong side of the fleece, weave the white strip through the slits in the center of the blue fleece and the lilac strips through the side slits.

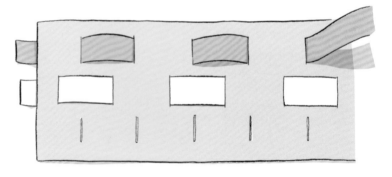

5 To make the pom-poms, cut four 3 x 10-in. (8 x 25-cm) strips of white fleece. Cut slits 1⅜ in. (3.5 cm) long along both long edges, spacing them ¼ in. (5 mm) apart. Roll up the fabric strips tightly, wrap a small length of embroidery floss (thread) around the middle of each one, and tie tightly in a knot. Trim the ends of the floss (thread).

6 Turn the scarf wrong side up. At each end of the scarf, cut a slit ¾ in. (2 cm) long in each woven fleece strip and tie the two ends in a knot.

7 Whipstitch (see page 113) the pom-poms to the corners of the woven scarf, using two strands of white embroidery floss (thread).

Super-warm and snuggly, this pretty scarf features a gorgeous flower pin and is the perfect accessory for a little girl dressed all in pink!

scarf with flower pin

Skill level: 1

You will need

Templates on page 115
½ yd (50 cm) coral fleece, 60 in. (150 cm) wide
4-in. (10-cm) square of lime-green felt
Light blue button, approx. ¼ in. (8 mm) in diameter
Pink button, ¾ in. (20 mm) in diameter
Dark pink polyester sewing thread
Brooch pin
Sewing machine
Basic sewing kit (see page 111)

Making the scarf

1 Cut two 6¼ x 40-in. (16 x 100-cm) strips of coral fleece. Place a small plate or bowl on the wrong side of each short end and draw around it with a permanent marker pen. Cut along your drawn lines to create rounded ends.

2 Pin the front and back of the scarf right sides together. Taking a ⅜-in. (1-cm) seam allowance, machine stitch around the edge, leaving a 3-in. (8-cm) gap at the bottom of the scarf for turning.

3 Trim the seam allowance to ⅛ in. (3 mm) to reduce the bulk, turn right side out, and whipstitch (see page 113) the opening closed.

Making the flower pin

4 Transfer the templates on page 115 onto card (see page 112) and cut out. Draw a 2½-in. (6-cm) diameter circle on card and cut out. This is the backing template. Place the flower and backing templates on the wrong side of the coral fleece, draw around them with a permanent marker pen, and cut out one of each shape. Place the leaf template on green felt, draw around it twice, and cut out.

5 Cover the fleece backing piece with fabric glue. Using the photo as a guide, coil up the flower, turn it wrong side up, and place the fleece backing over the center. Turn the flower right side up, position a blue and then a pink button in the center of the flower, and hand stitch in place.

6 Place the two leaf shapes together. Using two strands of pink sewing thread, work blanket stitch (see page 113) around the leaf. Turn the flower wrong side up, position the bottom point of the leaf in the center of the fleece backing, and hand stitch in place.

7 Cut out a ⅜ x ¾-in. (1 x 2-cm) rectangle of coral fleece and place it over the brooch bar back. Glue to the circular backing piece of the flower. Alternatively, hand stitch the brooch bar in place.

8 Overlap the rounded ends of the scarf and pin the fleece flower through both layers to join the front and back together.

These little animal badges would look great pinned onto children's clothing, accessories, or homewares or attached to a baby mobile. Or why not try making them bigger and turn them into pillows or toys?

animal badges

Skill level: 1

You will need

Templates on page 116

12-in. (30-cm) square each of light blue, lime-green, and orange fleece

6-in. (15-cm) square of mustard yellow felt

Dark blue, dark green, and brown polyester sewing thread

Two small pink buttons, ⅜ – ⅝ in. (1 – 1.5 cm) in diameter

Five googly eyes with eyelashes

Toy filling

Three brooch pins

Sewing machine

Basic sewing kit (see page 111)

| Transfer the animal templates on page 116 onto card (see page 112) and cut out.

Elephant

2 Place the elephant templates on the wrong side of the light blue fleece and draw around each one once with a permanent marker pen. Flip the body template over and draw around it once more, then cut out the shapes.

3 Pin the ear to the front of the body. Using a doubled length of dark blue thread, work a small backstitch (see page 112) all around the ear. Position two pink buttons on the front of the elephant and attach using fabric glue.

Frog

4 Place the frog template on the wrong side of the lime-green fleece, draw around it twice with a permanent marker pen, and cut out the shapes.

5 One the right side of one of the lime-green fleece pieces, using dark green thread, work two French knots (see page 113) for the nose and a line of backstitch (see page 112) for a smiley mouth.

Lion

6 Place the lion outer template on the wrong side of the orange fleece, draw around it twice, and cut out the shapes. Place the lion inner template on the mustard yellow felt, draw around it once with a permanent marker pen, and cut out the shape.

7 On the lion inner fleece piece, using a doubled length of brown thread, work French knots for the whiskers, a backstitch for the mouth, and a satin stitch (see page 113) for the nose. Place one of the lion outer fleece pieces right side up, then pin the lion inner felt piece in the center. Using a doubled length of brown thread, work a small backstitch all around the edge of the inner piece.

Assembling the badges

8 Place the front and back pieces of each animal together, so that the right sides are facing outward, and pin together in the center. Using two strands of the same thread used for the main design, work a small backstitch all around each shape, stitching through both layers and leaving a small gap for filling. Fill each animal badge with toy stuffing, then backstitch the opening closed.

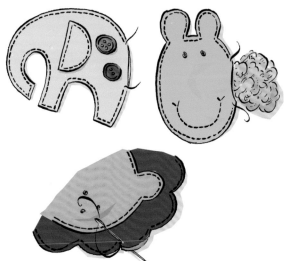

9 Using the photo as a guide, place googly eyes on the front of the fleece animals and attach using fabric glue.

10 Cover the base of the brooch pin bars with fabric glue and place one in the center back of each animal; alternatively, stitch in place by hand.

Whether you hang them on the wall or leave them freestanding, these simple letters will give your child's bedroom a unique, personalized finishing touch.

letter wall art

Skill level: 1

You will need

20-in. (50-cm) square of fleece fabric for each letter

Polyester sewing threads that match the color/pattern of fleece fabric

Toy filling

Sewing machine

Basic sewing kit (see page 111)

1 Choose a font, type the letters you want in your word-processing or desktop-publishing program, then change the font size so that each letter is approx. 11 in. (28 cm) tall and fits on a piece of letter-size (A4) paper. Print out, then transfer to card to make a template (see page 112).

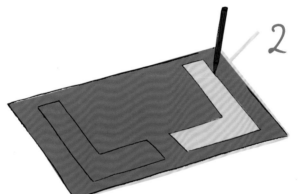

2 Place each template in turn on the wrong side of the appropriate fleece fabric and draw around it with a permanent marker pen. Flip the template over, then draw around it once more. Cut out all the pieces.

3 Place the fleece pieces right sides together in pairs and pin around to hold. Taking a ¼-in. (5-mm) seam allowance, machine stitch all the way around the letters, leaving a 4-in. (10-cm) gap at one side for turning. For letters A, B, P, and R, which have small internal holes, leave the internal holes unstitched. For letters D, Q, and O, which have large internal holes, machine stitch three-quarters of the way around the inside open edges, leaving a 4-in. (10-cm) gap for turning opposite the open side edge.

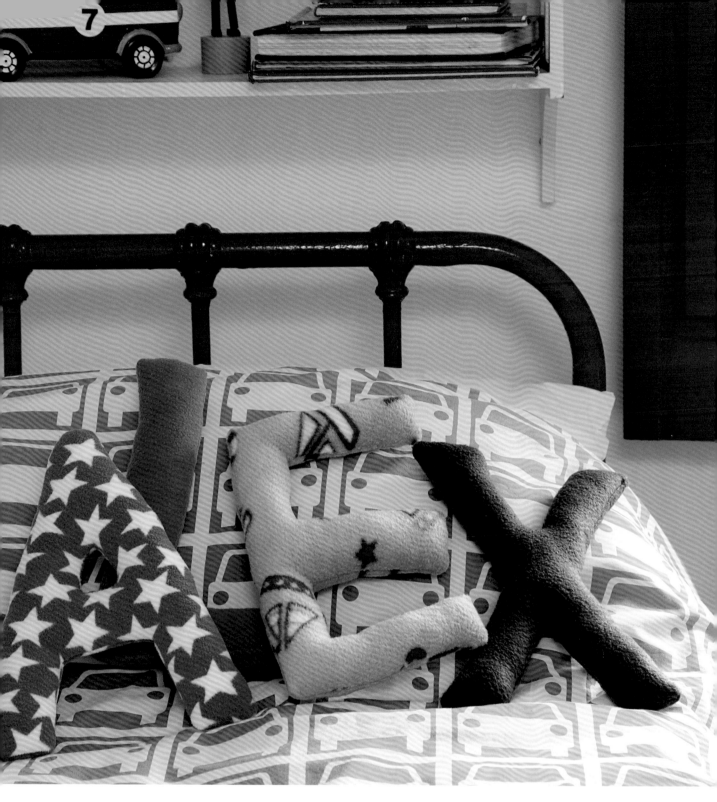

4 Trim all seam allowances to ⅛ in. (3 mm) to reduce the
bulk and turn right side out. Fill with toy filling. Using
a doubled length of matching sewing thread,
whipstitch (see page 113) any openings closed.

Chapter 3

For the home

There are so many stylish items that you can make for your home using fleece—from tea cozies to decorative pillows and hoop art, the possibilities are endless. After flicking through this chapter, you'll soon be sewing up lots of fleecy wares for your home!

Make your very own appliquéd love heart hug for your favorite mug following these simple steps. It's perfect for keeping your tea or coffee hot.

mug cozy

Skill level: 1

You will need

Templates on page 124
20-in. (50-cm) square of
   fusible bonding web
10 in. (25 cm) each of
   gray and lilac fleece,
   20 in. (50 cm) wide
6-in. (15-cm) square
   each of mint green
   and coral felt
Coral button, approx.
   ⅝ in. (1.5 cm) in
   diameter
Coral polyester sewing
   thread
Sewing machine
Basic sewing kit (see
   page 111)

1   Transfer the templates on page 124 onto card (see page 112) and cut out. (Depending on the size of your mug, you may need to shorten or lengthen the front and back cozy templates.) Draw around the front cozy and the heart templates on the paper side of a piece of fusible bonding web.

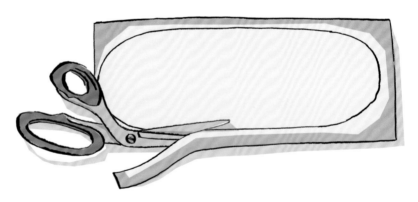

2   Roughly cut out the front cozy shape from the bonding web. Following the manufacturer's instructions, apply the shape to the wrong side of the gray fleece across the grainline (see page 114). Cut out the cozy shape along your drawn line and peel off the backing paper.

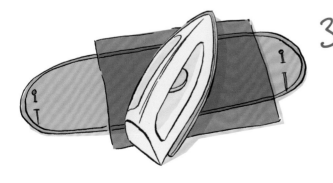

3   Place the back cozy template on the wrong side of the lilac fleece across the grainline, draw around it, and cut out the shape. With wrong sides together, place the back fleece cozy on the front fleece cozy, and pin at either side to hold. Press with an iron on a low setting for a few seconds to fuse the fabrics together (see page 114).

4 Roughly cut out all the heart shapes that you drew on the bonding web. Following the manufacturer's instructions, apply the large heart to mint green felt, the medium heart to the wrong side of the lilac fleece, and the small heart to the coral felt. Cut out, then peel off all the backing papers.

5 Using the photo as a guide, apply the hearts to the front of the cozy. Machine stitch around the edges of the heart shapes and cozy.

6 Sew a coral button to the right-hand outer side of the cozy.

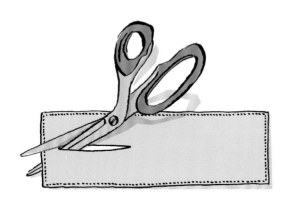

7 To make the buttonhole band, cut two 2 x ¾-in. (5 x 2-cm) rectangles of mint green felt. Place one on top of the other and straight stitch around the edges. Make a small snip through both layers near the left-hand side, big enough for the button to go through.

8 Place the mint green felt under the left-hand edge of the cozy, pin to hold, and machine stitch in place.

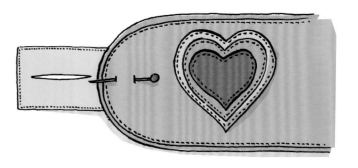

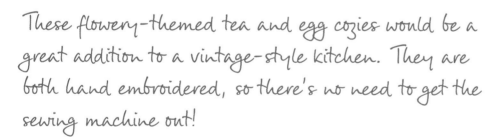

Skill level: 1

You will need
For one tea cozy and
   two matching egg
   cozies, you will need:
Templates on page 122
1 yd (1 m) light blue
   fleece
6-in. (15-cm) square of
   pink felt fabric
20-in. (50-cm) square
   of yellow fleece
9 orange buttons, ⅜ in.
   (1 cm) in diameter
   for the tea cozy and
   1 for each egg cozy
Orange embroidery
   floss (thread)
Basic sewing kit (see
   page 111)

These flowery-themed tea and egg cozies would be a great addition to a vintage-style kitchen. They are both hand embroidered, so there's no need to get the sewing machine out!

kitchen cozies

To make the tea cozy

1 | Transfer the tea cozy template on page 122 onto card (see page 112) and cut out. Place on the wrong side of the light blue fleece (on the grainline). Using a permanent marker pen, draw around it four times. Cut out all the pieces.

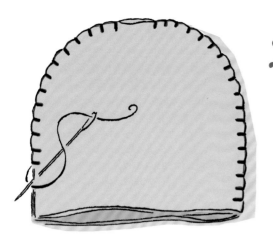

2 Place the pieces together in pairs, with the wrong sides together. Place these pairs together and pin around the sides to hold. Stitching through all four layers, work a blanket stitch (see page 113) around the curved outside edge in orange embroidery floss (thread), leaving the straight bottom edge open and a small gap at the top for the felt loop.

3 Blanket stitch the straight bottom edges of the front and back of the cozy, stitching through two layers of fleece.

4 Cut out one 2 x ⅝-in. (5 x 1.5-cm) strip of pink felt to make a loop. Place inside the unstitched gap at the center top of the tea cozy and attach using fabric glue.

5 Draw a circle 2⅜ in. (6 cm) in diameter on card and cut out. Place on the wrong side of the yellow fleece, draw around it nine times, and cut out.

6 Fold each fleece circle in half, wrong sides together. Pin to hold, and make five small wedge-shaped snips evenly around the semicircles to make petals. Open out the circle, then cut a curved shape around the top of each petal.

7 Using the photo as a guide,
position the yellow fleece
flowers on the tea cozy and
attach using fabric glue. Glue
an orange button to the center
of each flower.

To make the egg cozies

The egg cosies are made in exactly
the same way as the tea cozy, using
the template on page 122.

A fantastic "no-sew" project, these fleecy coaster designs are sure to brighten up your living room!

coiled coasters

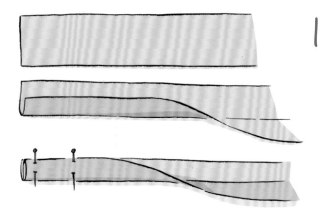

1 For each coaster, cut a strip of fleece in your chosen color measuring 2½ x 44 in. (6 x 112 cm). Place right side down on your work surface, fold the long edges of each strip in to the center, then fold in half lengthwise. Pin across to hold the folds together.

Skill level: 1

You will need
2½ x 44 in. (6 x 112 cm) each of denim blue, pink, mustard yellow, and red fleece
Basic sewing kit (see page 111)

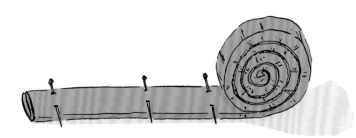

2 Lay the strip on a flat surface and roll it up tightly, removing the pins as you go. Secure the end of the strip in place with a small amount of fabric glue.

3 Draw a circle 3½ in. (9 cm) in diameter on card and cut out. Place it on the wrong side of each shade of fleece in turn, draw around it once, and cut out. This will be the backing circle.

4 Turn the coaster over, cover it with fabric glue, and place the backing circle on the bottom of the coaster.

Filled with a mixture of play sand and toy stuffing, this pretty house design would be perfect for either propping open a door or supporting a tumbling pile of books.

house doorstop

Skill level: 2

You will need
Templates on page 123
40 x 60 in. (100 x
    150 cm) fluffy
    cream fleece
12-in. (30-cm) square
    of denim blue fleece
6-in. (15-cm) square
    each of coral pink
    and white fleece
Pink, green, blue, and
    cream polyester
    sewing threads
Small bag of play sand
Toy filling
Sewing machine
Darning foot
Basic sewing kit (see
    page 111)

1 Transfer the house template on page 123 onto card (see page 112) and cut out. Place the template on the wrong side of the cream fleece (on the grainline). Using a permanent marker pen, draw around it twice, and cut out both pieces.

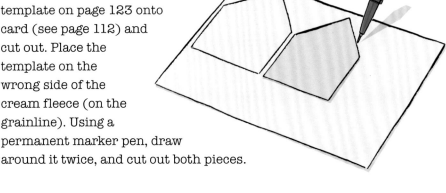

2 Cut out one 5 x 21¼-in. (12.5 x 54-cm) piece of cream fleece for the sides, one 5 x 7-in. (12.5 x 17.5-cm) piece for the base, and two 6¼ x 11-in. (16 x 28-cm) pieces of denim blue fleece for the roof.

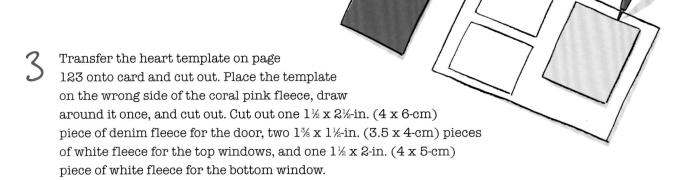

3 Transfer the heart template on page 123 onto card and cut out. Place the template on the wrong side of the coral pink fleece, draw around it once, and cut out. Cut out one 1½ x 2½-in. (4 x 6-cm) piece of denim fleece for the door, two 1⅜ x 1½-in. (3.5 x 4-cm) pieces of white fleece for the top windows, and one 1½ x 2-in. (4 x 5-cm) piece of white fleece for the bottom window.

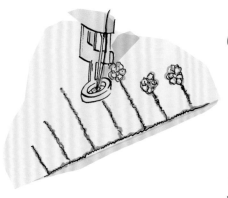

4 Fit a darning foot to your sewing machine and free-motion stitch (see page 113) flowers and grass along the bottom edge, using pink and green sewing threads.

5 Using the photo as a guide, pin the windows, door, and heart onto the front fleece house piece. Change to blue thread and free-motion stitch around each piece.

Assembling the house

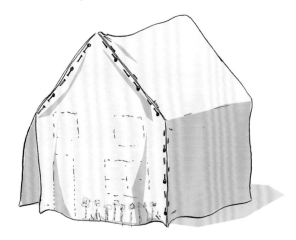

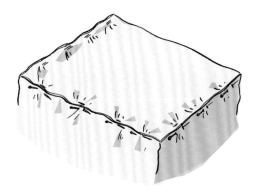

6 Change back to a normal sewing machine foot and cream thread. With right sides together, pin the side strip and house front sections together. Machine stitch, taking a ⅜-in. (1-cm) seam allowance.

7 Pin the base to the bottom edges of the side and front house sections. Machine stitch, taking a ⅜-in. (1-cm) seam allowance.

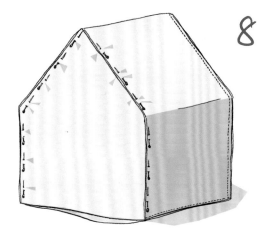

8 Pin the back house section onto the side piece. Machine stitch, taking a ⅜-in. (1-cm) seam allowance, leaving the bottom edge open.

9 Trim all seam allowances to ⅛ in. (3 mm) to reduce the bulk and turn right side out. Insert the bag of play sand, then fill up the rest of house with toy filling, and whipstitch (see page 113) the opening closed.

10 Place the roof pieces wrong sides together and pin to hold. Cut five small wedge-shaped snips evenly along each short edge, then round off each section into a curved, scallop shape.

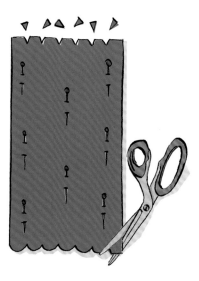

11 Fold the roof in half widthwise. Change to an embroidery foot and blue thread. Free-motion stitch along the folded edge, open the back out, and stitch along the long side edges and curved roof edges. Remove the pins. To create a tiled-roof effect, free-motion stitch scallop shapes all the way across the roof, using the photo as a guide.

12 Place the roof on top of the stitched house and attach using fabric glue.

13 Cut out a 10 x 4-in. (25 x 10-cm) piece of coral pink fleece for the handle. Fold it in half lengthwise, with right sides together, and pin along the long unfolded edge. Straight stitch along, leaving the sides open. Turn right side out.

14 Tuck in the raw short ends by ⅜ in. (1 cm). Using the photo as a guide, position the handle on the top of the roof, pin at the ends to hold, and whipstitch in place.

An ideal project for those long winter months, this draft excluder is so easy to make and requires no sewing. All you have to do is tie the fringes together—and voilà! You have the perfect project for keeping out cold drafts!

draft excluder

Skill level: 1

You will need

24 in. (60 cm) patterned blue-and-white star fleece, 60 in. (150 cm) wide

Vanishing marker pen

Toy filling

Long ruler

Basic sewing kit (see page 111)

1 Lay the fleece on a flat surface and cut out two 12 x 40-in. (30 x 100-cm) strips. Place them wrong sides together, and pin together along the center.

2 Cut a piece of card 2 in. (5 cm) square. Lay the fleece on a flat surface. Place the card square on each corner of the fleece in turn and draw around it with a vanishing fabric marker pen. Cut away the square corner sections, cutting through both layers of fleece.

3 Place a long ruler on the front fleece piece, aligning it with the inside edges of the cut-out corners. Using a vanishing fabric marker pen, draw a line all around.

4 Cut 2-in. (5-cm) slits ⅜ in. (1 cm) apart all around the fleece, cutting up to the lines you drew in step 3 and making sure you cut through both thicknesses of fleece.

5 To attach the top piece of fleece to the bottom piece, tie the fringes together in a double knot. Work three-quarters of the way around, then remove the pins. Stuff with toy filling, then finish knotting the fringes together.

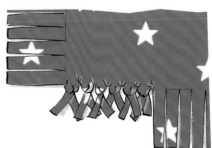

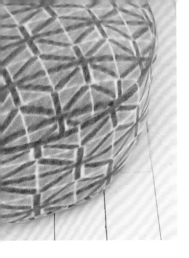

Super soft and extremely comfortable to sit on, this fleecy stool is sure to be a favorite in your home. Why not choose a stylish patterned fleece to match your décor?

comfy pouffe

Skill level: 1

You will need

40 x 60 in. (100 x 150 cm) patterned fleece
Matching polyester sewing thread
Toy filling
Sewing machine
Basic sewing kit (see page 111)

1 Draw a circle 20½ in. (52 cm) in diameter on card and cut out. Place on the wrong side of the patterned fleece (on the grainline). Using a permanent marker pen, draw around it twice. Cut out both pieces.

2 Cut out a 7 x 65-in. (18 x 165-cm) strip of patterned fleece. Fold it in half widthwise, right sides together, and machine stitch down the short side edges ⅜ in. (1 cm) in from the edge to form a loop.

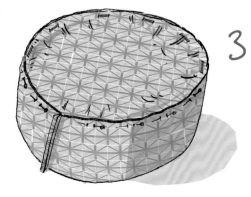

3 Pin the raw edge of the loop to the wrong side of one fleece circle. Taking a ⅜-in. (1-cm) seam allowance, machine stitch all the way around.

4 Pin the other side of the loop to the wrong side of the remaining fleece circle. Taking a ⅜-in. (1-cm) seam allowance, machine stitch all the way around, leaving a 4-in. (10-cm) gap for turning. Trim all the seam allowances to ⅛ in. (3 mm) to reduce the bulk.

5   Turn right side out, fill with toy filling, and whipstitch (see page 113) the opening closed.

Featuring pretty trims of pom-pom, ric-rac, and sequins, this adorable heart pillow would make a beautiful addition to a little girl's chair or bed.

heart pillow

Skill level: 1

You will need
Template on page 126
20-in. (50-cm) square of
    fuchsia pink fleece
1 yd (1 m) medium pink
    pom-pom trim, ⅝ in.
    (18 mm) wide
20 in. (50 cm) turquoise
    sequin trim, ¼ in.
    (6 mm) wide
1 yd (1 m) lilac ric-rac
    trim, approx. ¼ in.
    (8 mm) wide
1 yd (1 m) turquoise
    pom-pom trim, 1⅜ in.
    (34 mm) wide
Fuchsia pink polyester
    sewing thread
Toy filling
Sewing machine
Basic sewing kit (see
    page 111)

1 Transfer the template on page 126 onto card (see page 112) and cut out. Place the template on the wrong side of the fuchsia pink fleece (on the grainline). Using a permanent marker pen, draw around it twice, and cut out both pieces.

2 Using the photo as a guide, stretch the pink pom-pom, turquoise sequin, and lilac ric-rac trims across the right side of one fleece heart. Cut to the required length and attach each one using fabric glue.

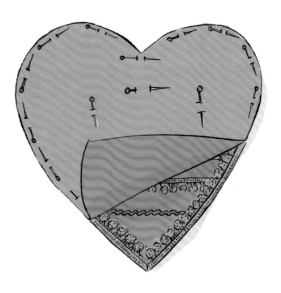

3 Place the turquoise pom-pom trim around the edge of the decorated front piece, with the pom-poms facing inward. Pin the back piece right side down on top. Place pins around the center of the pillow to hold. Starting at the bottom corner, machine stitch all the way around, working as close to the edge as possible and leaving a 3-in. (8-cm) gap at the side edge for turning.

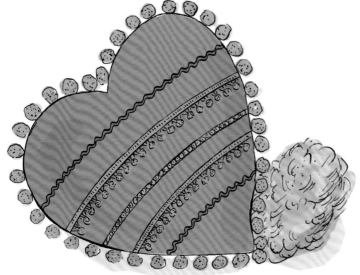

4 Turn right side out and fill with toy filling. Tuck the pom-pom tape inside the unstitched gap, fold in the fleece fabric, and pin to hold. Work a small backstitch (see page 112) across the open edge and the pom-pom tape to close the opening.

Drape this fleecy blanket over your bed and give yourself some extra warmth in the wintertime! The fringed edge, with its tiny jewel-like bursts of color, adds a fun, happy touch.

fringed blanket

Skill level: 1

You will need
40 x 60 in. (100 x 150 cm) white fleece
Vanishing marker pen
6-in. (15-cm) squares of felt in different colors
Long ruler
Basic sewing kit (see page 111)

1 Trim off any selvages around the edges of the fleece. Lay the fleece on a flat surface. Cut a piece of card 2¾ in. (7 cm) square. Place the card square on each corner of the fleece in turn and draw around it with a vanishing fabric marker pen. Cut away the square corner sections.

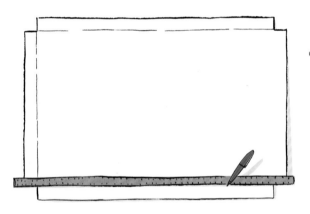

2 Place a long ruler on the fleece, aligning it with the inside edges of the cut-out corners. Using a vanishing marker pen, draw a line all around.

3 Cut 2¾-in. (7-cm) slits ¼ in. (5 mm) apart all around the fleece, cutting up to the lines you drew in step 2.

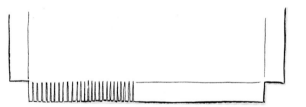

4 Cut the colored felt into 4 x ¾-in. (10 x 2-cm) strips. Gather 6–7 fringes together and tie a felt strip tightly around each one. Continue all the way around, alternating the colors of felt. Cut away the excess felt.

This sweet little hoop-art design is so simple and quick to make, and would look lovely hung on a child's bedroom wall. It is based on a traditional patchwork design known as Suffolk puffs.

hoop art

Skill level: 1

You will need
Template on page 116
12-in. (30-cm) square
    each of lilac, pink
    polka-dot, and green
    fleece
Vanishing fabric
    marker pen
Dark green and pink
    polyester sewing
    threads
Three small purple
    buttons
8¼-in. (21-cm)
    embroidery hoop
Sewing machine
Darning foot
Basic sewing kit (see
    page 111)

1 Place the fleece wrong side up in the embroidery hoop. Turn the hoop over so that the fleece fabric can lie flat on the needle plate of your machine and you're stitching on the right side of the fabric.

2 Using the photo as a guide, draw flower stems on the fleece with a vanishing fabric marker pen. Fit a darning foot to your machine and thread it with green sewing thread. Free-motion stitch (see page 113) over the marked-out stems.

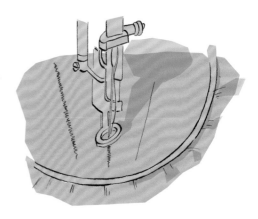

3 Transfer the leaf template on page 116 onto card (see page 112) and cut it out. Place the template on the wrong side of the green fleece, draw around it four times, and cut out the shapes.

4 Draw three circles 3½ in. (8.5 cm) in diameter on the wrong side of the polka-dot fleece and cut out.

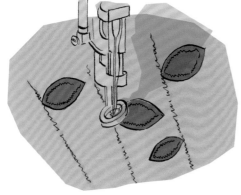

5 Using the photo as a guide, place the leaves on the lilac fleece background. Pin in place, then free-motion stitch around the edge of each leaf shape.

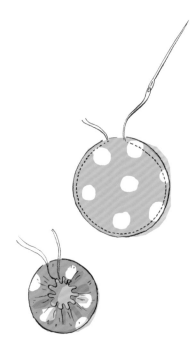

6   With the wrong sides of the fleece circles facing you, work a line of running stitch all around ¼ in. (5 mm) from the edge, using a doubled length of pink thread. Pull both thread ends to gather the fabric and tie the threads tightly in a knot.

7   Place the Suffolk puff flowers on the fleece background and attach using fabric glue. Stitch a purple button to the center of each flower through the fleece background.

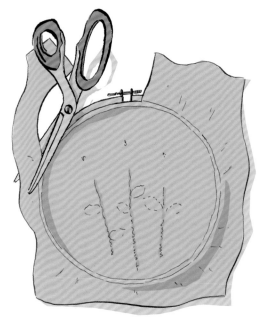

8   Remove the fabric from the hoop, turn it over, and stretch it over the inner hoop, with the design facing upward. Place the outer hoop on top and screw it in place. Cut away the excess fleece fabric to create a neat edge.

It takes no time at all to stitch this stylish hot-water bottle cover.
The wrap-around scarf makes it extra cozy!

retro bottle cozy

Skill level: 1

You will need
Templates on pages
   124–125
40 x 60 in. (100 x
   150 cm) patterned
   fleece
20-in. (50-cm) square
   of orange fleece
Sewing thread
Sewing machine
Basic sewing kit (see
   page 111)

| Transfer the
templates on pages
124–125 onto card
(see page 112) and
cut out. Place all
three templates on
the wrong side of the
patterned fleece (on
the grainline). Using
a permanent marker
pen, draw around
each one once. Cut
out all the pieces.

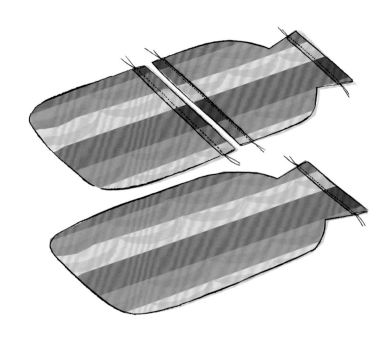

2 Fold the top and bottom edges of the upper front piece over to the wrong side by ⅜ in. (1 cm) and machine stitch across. Fold the top edges of the lower front piece and the back piece over to the wrong side by ⅜ in. (1 cm) and machine stitch across.

3 Place the front and back pieces right sides together and pin to hold, overlapping the lower front piece on top of the upper front piece. Taking a ⅜-in. (1-cm) seam allowance, machine stitch all the way around the bottom and side edges. Trim the seam allowance and turn right side out.

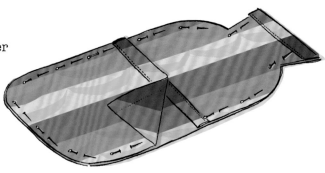

4 Cut out two 2½ x 20-in. (6 x 50-cm) strips of orange fleece. Cut slits 2 in. (5 cm) long, ¼ in. (5 mm) apart, along both short edges of each piece.

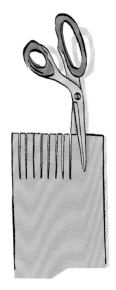

5 Place the fleece scarves together, wrap them around the hot water bottle cover, and tie in a knot to secure.

tools and techniques

Equipment

The projects in this book can all be made with very little equipment. A sewing machine is essential, but it can be a very simple model. Even the most basic of modern sewing machines offers a variety of stitches.

Basic stitches such as a straight or zig-zag can be achieved using the standard presser foot (also referred to as an all-purpose foot or zig-zag foot), which has a wide needle hole that fits the widest stitch on your sewing machine. Most of the projects contained in this book require you to use one of these, but there are some that feature free-motion embroidery detail, in which case you will need a darning foot (also referred to as an embroidery foot). A darning foot will either maintain a "C" or an "O" shape and will have a clear plastic or metal base. Most machines will have a darning plate (or a function to lower the feed dogs for machine embroidery) but will not come with a darning foot. When purchasing always refer to manufacturer's guidelines and make sure you buy one that is compatible with your machine.

Basic sewing kit

In addition, you'll need the following:
- Tracing paper and pencil for tracing all of the project templates
- Ruler and tape measure for making up projects that don't require a pattern and for checking dimensions
- White card for making up templates for drawing around on fleece
- Permanent marker pen for drawing around templates and patterns
- Fabric and paper scissors for cutting into fleece fabric and creating card templates
- Dressmaking pins for holding together fleece fabrics before sewing in the machine
- Hand-sewing needles for attaching embellishments and whipstitching projects
- Embroidery needles for working embroidery stitches such as crosses and blanket stitches
- Fabric glue for attaching fleece shapes and backings
- Iron for bonding fabrics together

Fleece fabrics

Fleece is a synthetic fabric made up of man-made fibers that have twisted together. Not only does it come in a wide variety of solid colors and vibrant patterns, but there are also different types of fleece. Anti-pill polar fleece, which I have used for all the projects, is the most popular one. It can be bought from most fabric stores, and is normally sold in two different lengths: 20 x 60 in. (50 x 150 cm) and 40 x 60 in. (100 x 150 cm).

Polar fleece has many qualities. It is snugly-soft, lightweight, and extremely comfortable to wear, and even though it's manufactured from 100 percent polyester, it's extremely cozy. However the most amazing of its properties is its ability to retain its insulation when wet, which makes it a very popular textile choice for outdoor wear.

Sewing fleece by machine

Fleece is relatively easy to sew in the machine but its stretchiness and thickness can sometimes cause problems. There are a few things you can do to make it an easy and enjoyable process.
- Pin the fabric as much as possible on the area you wish to sew across/around.
- Select a slightly larger stitch length than normal.
- Reduce your tension or loosen your presser foot.
- Sew very slowly and carefully.
- Use a thick sewing machine needle such as a standard point/universal 90/14. The larger the eye and the stronger the needle, the better.
- Use 100 percent polyester sewing thread. Cotton threads often do not work as well and can end up breaking.

Hand sewing and embroidering into fleece

When hand stitching into fleece you don't often encounter too many problems but its best to always use good quality threads and sharp pointed needles.

Using templates

The best way to trace the templates contained in this book is to use tracing paper and a soft-leaded pencil. Start by tracing the required design onto tracing paper. Then turn the paper over, place it on top of white card, and rub over your lines with your pencil. The image will then be transferred and you will be able to cut out your shape.

The templates on pages 115–127 specify how much you need to enlarge motifs in order for your project to be the same size as the ones I made, but it's worth knowing how to enlarge or reduce motifs to any size you want.

Enlarging motifs

First, decide how big you want the pattern or motif to be. Let's say that you want a particular shape to be 10 in. (25 cm) tall.

Then measure the template that you are working from—5 in. (12.5 cm) tall, for example.

Take the size that you want the pattern or motif to be (10 in./25 cm) and divide it by the actual size of the template (5 in./12.5 cm). Multiply that figure by 100 and you get 200—so you need to enlarge the motif on a photocopier to 200 percent.

Reducing motifs

If you want a motif on the finished piece to be smaller than the template in the book, the process is exactly the same. For example, if the pattern is 5 in. (12.5 cm) tall and you want the motif to be 3 in. (7.5 cm) tall, divide 3 in./7.5 cm by the actual size of the template (5 in./12.5 cm) and multiply by 100, which gives you a figure of 60. So the figure that you need to key in on the photocopier is 60 percent.

Transferring patterns and motifs onto fleece

To transfer designs onto fleece, place your template/pattern on the wrong side of the fleece fabric. It's a much flatter surface than the right side, and so is easier to draw on.

To determine the grainline, gently pull and stretch the edge of the fabric that has been cut across from selvage to selvage. Fleece stretches in only one direction, so ALWAYS make sure that you place the pattern/template on the fabric grainline (along the stretch). It is particularly important to do so when making projects such as hats, headbands, socks, and jackets.

Always use a permanent marker pen when drawing around the template/pattern, holding it in place with one hand while you carefully draw around it. Alternatively, pin it in place.

Hand stitches

There are literally hundreds of decorative hand stitches that you can use to embellish your projects. Here are some of the most useful.

Running stitch

Work from right to left. Secure the thread with a couple of small stitches, and then make several small stitches by bringing the needle up and back down through the fabric several times along the stitching line. Pull the needle through and repeat. Try to keep the stitches and the spaces between them the same size.

Backstitch

Work from right to left. Bring the needle up from the back of the fabric, one stitch length to the left of the end of the stitching line. Insert it one stitch length to the right, at the very end of the stitching line, and bring it up again one stitch length in front of the point from which it first emerged. Pull the thread through. To begin the next stitch, insert the needle at the left-hand end of the previous stitch. Continue to the end.

French knot

Bring the needle up from the back of the fabric to the front. Wrap the thread two or three times around the tip of the needle, then reinsert the needle at the point where it first emerged, holding the wrapped threads with the thumbnail of your non-stitching hand, and pull the needle all the way through. The wraps will form a knot on the surface of the fabric.

Satin stitch

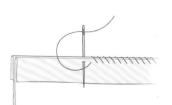

This is a "filling" stitch that is useful for motifs such as flower petals and leaves. Work from left to right. Draw the shape on the fabric, then work straight stitches across it, coming up at A and down at B, then up at C and down at D and so on. Place the stitches next to each other, so that no fabric can be seen between them. You can also work a row of backstitch around the edge to define the outline more clearly.

Blanket stitch

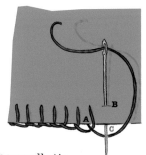

Bring the needle out at the edge of the fabric at A. Insert the needle at B, to the right and above the edge, then bring it down to the edge of the fabric at C, keeping the thread under the needle tip. Draw the needle through to form a looped stitch.

Whipstitch

Whipstitch is generally used to stitch two finished edges together. Working from right to left, insert the needle at right angles through the edges, picking up one or two threads from the back and then from the front edge; draw the thread through. Insert the needle in the back edge to the left of the first stitch and bring it out through the front edge. Continue in this way until the two edges are joined. Slanted stitches are produced, which can be short or long, depending on how close together your stitches are.

Free-motion machine embroidery

With free-motion embroidery (sometimes referred to as "drawing with stitch"), you have the freedom to move the stitching anywhere you want, in any direction, and create any design.

When creating free-stitched detail on fleece, you will need to drop your feed dogs (or fit a darning plate if your machine does not have this function); this will stop the machine from feeding your fabric. You will also need to adjust your tension to 0 or 1. With your fabric stretched tightly around a wooden embroidery hoop and placed under the foot in the machine, slowly and carefully move your embroidery hoop with both hands (as if it is a steering wheel)

while stitching on top of the fleece to create your design. Keep moving slowly until you get a rhythm going. You can either create one line of stitching or go over it a few times to create a more decorative stitched pattern. Always bear in mind that this type of stitching will never look straight and perfect, but that's the idea. It should look squiggly, sometimes with curved lines going here, there, and everywhere.

If you are new to free-motion embroidery, it is useful to get really comfortable with the technique before you jump into your project. Practice different stitches on a scrap piece of fabric such as swirls, zig-zags, and straight lines. Also practice trying to match the speed of your hands to the speed of the needle. You determine the stitch length this way. When it comes to working on your project, you might want to mark the areas you need to stitch around/across using a vanishing marker pen, which can make the process a little easier for beginners.

Appliqué using bonding web

Appliqué is a very versatile method of applying small pieces of fabric to a background fabric to create decorative effects. One of the simplest methods is using paper-backed adhesive web. This paper-backed adhesive web is ironed onto your fabric, which melts the glue, allowing you to stick fabric pieces together. Bear in mind that fleece has a very low melting point, so never press the iron onto the fabric for too long and always make sure you place a cover on top of the fleece to avoid marking or burning it.

1   Cut a piece of contrast fabric and paper-backed adhesive web (available from good haberdashery stores) large enough to fit the motif that you want to appliqué. Using a hot iron and following the manufacturer's instructions, stick the adhesive web to the wrong side of your contrast fabric.

2   Trace your motif onto the paper side of the adhesive web and cut out the shape.

3   Remove the paper backing from your motif and position it on your main fabric. Place a cover on top of the motif and fabric before ironing. Using an iron on a low setting and following the manufacturer's instructions, stick the motif in place.

4   Set your sewing machine to a medium stitch width. Using a close zig-zag stitch, stitch around the edges of your motif, enclosing all the raw edges. You can use either a matching or a contrasting thread.

Box stitching

As the name suggests, box stitching simply means stitching in a box shape with a cross in the center. It is used wherever a strong, reinforced join is needed—for example, when attaching straps to bags.

Starting at one edge, machine stitch across the strap in a square, finishing with the needle down. Pivot the work around the needle and stitch diagonally across the square to the opposite corner, then along the side of the square over the first line of stitching, and finally diagonally across the square to the opposite corner.

templates

All the templates required to make the projects are provided here. The
templates on pages 115–116 are full-size templates and can be traced off
the page, while those on pages 117–127 are half-size templates and will
need to be photocopied at 200% to print them out at the correct size.
Where there are three sizes of template, three outlines have been provided.
All templates include the seam allowance.

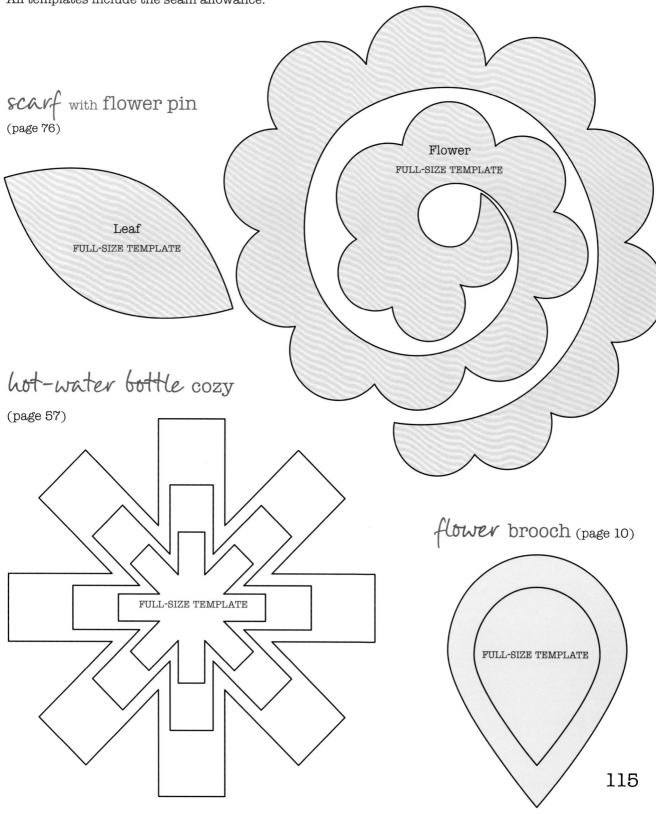

scarf with **flower pin**
(page 76)

Leaf
FULL-SIZE TEMPLATE

Flower
FULL-SIZE TEMPLATE

hot-water bottle cozy
(page 57)

FULL-SIZE TEMPLATE

flower brooch (page 10)

FULL-SIZE TEMPLATE

animal badges (page 79)

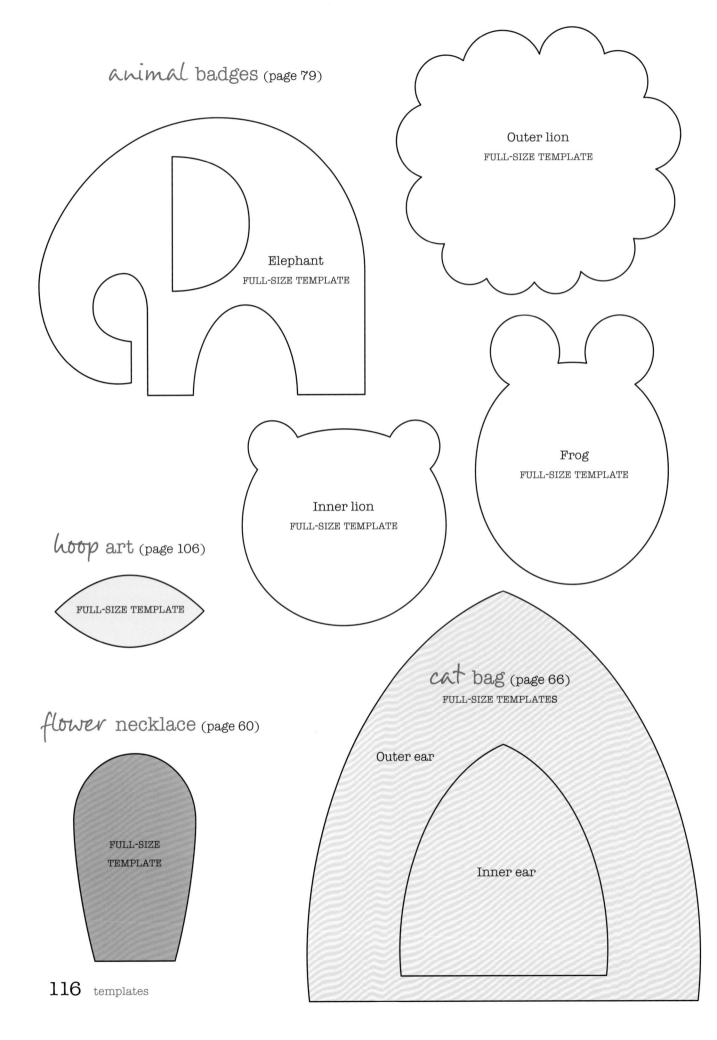

Elephant
FULL-SIZE TEMPLATE

Outer lion
FULL-SIZE TEMPLATE

Frog
FULL-SIZE TEMPLATE

Inner lion
FULL-SIZE TEMPLATE

hoop art (page 106)

FULL-SIZE TEMPLATE

flower necklace (page 60)

FULL-SIZE
TEMPLATE

cat bag (page 66)
FULL-SIZE TEMPLATES

Outer ear

Inner ear

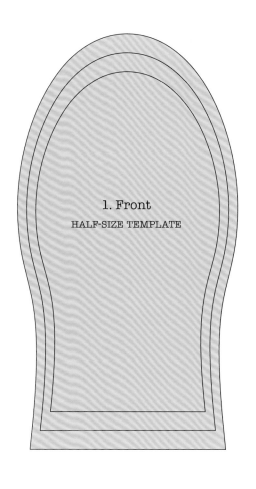

1. Front
HALF-SIZE TEMPLATE

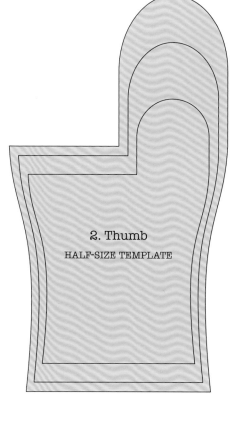

2. Thumb
HALF-SIZE TEMPLATE

Spike
HALF-SIZE TEMPLATE

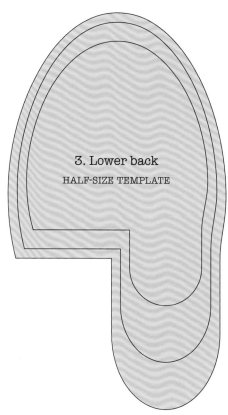

3. Lower back
HALF-SIZE TEMPLATE

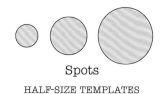

Spots
HALF-SIZE TEMPLATES

spotty boot warmers (page 13)

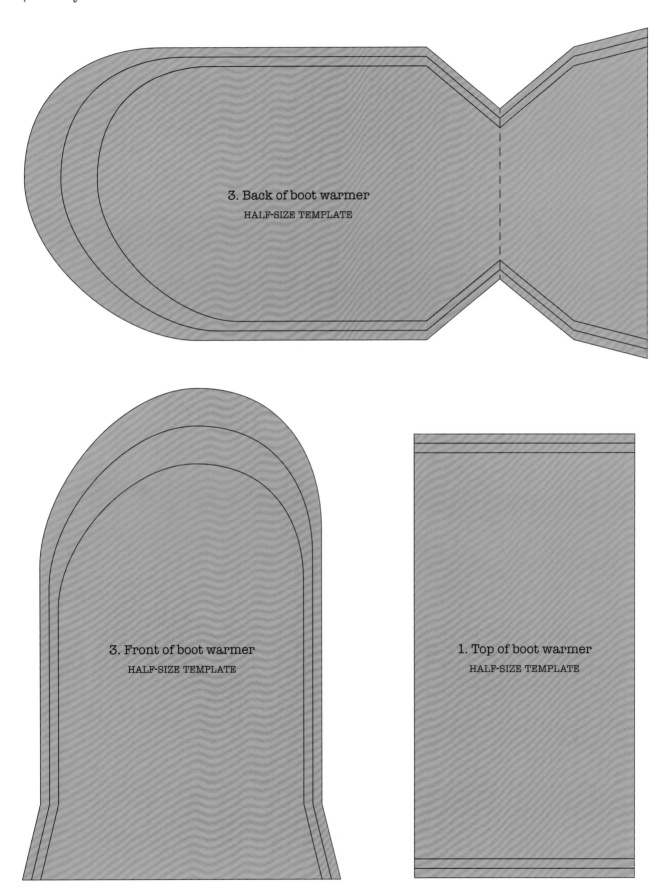

3. Back of boot warmer
HALF-SIZE TEMPLATE

3. Front of boot warmer
HALF-SIZE TEMPLATE

1. Top of boot warmer
HALF-SIZE TEMPLATE

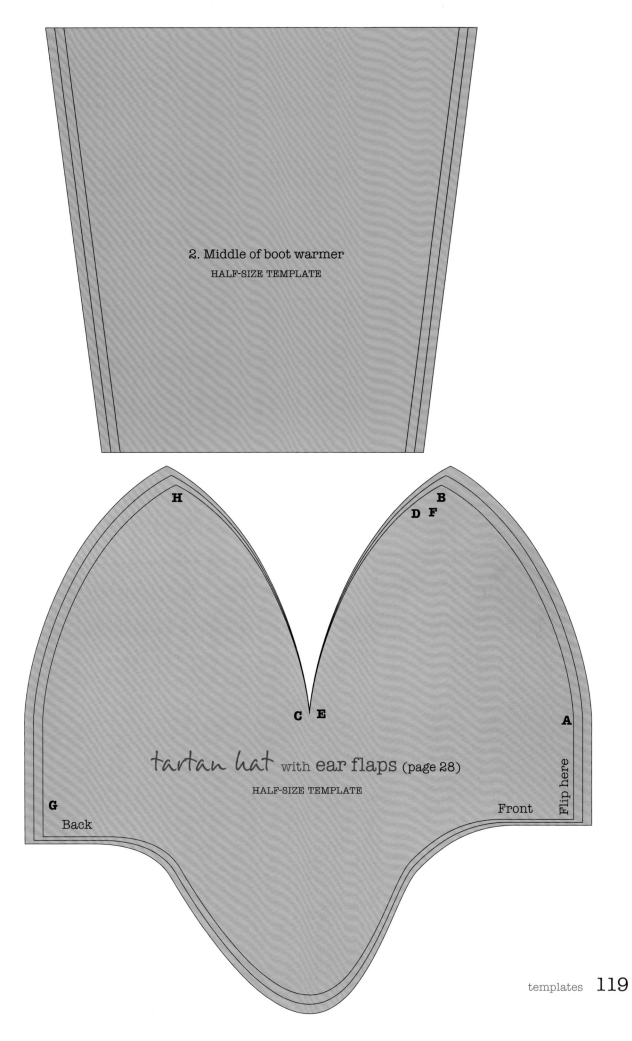

2. Middle of boot warmer
HALF-SIZE TEMPLATE

H

B

D F

C E

A

tartan hat with *ear flaps* (page 28)
HALF-SIZE TEMPLATE

G

Front

Flip here

Back

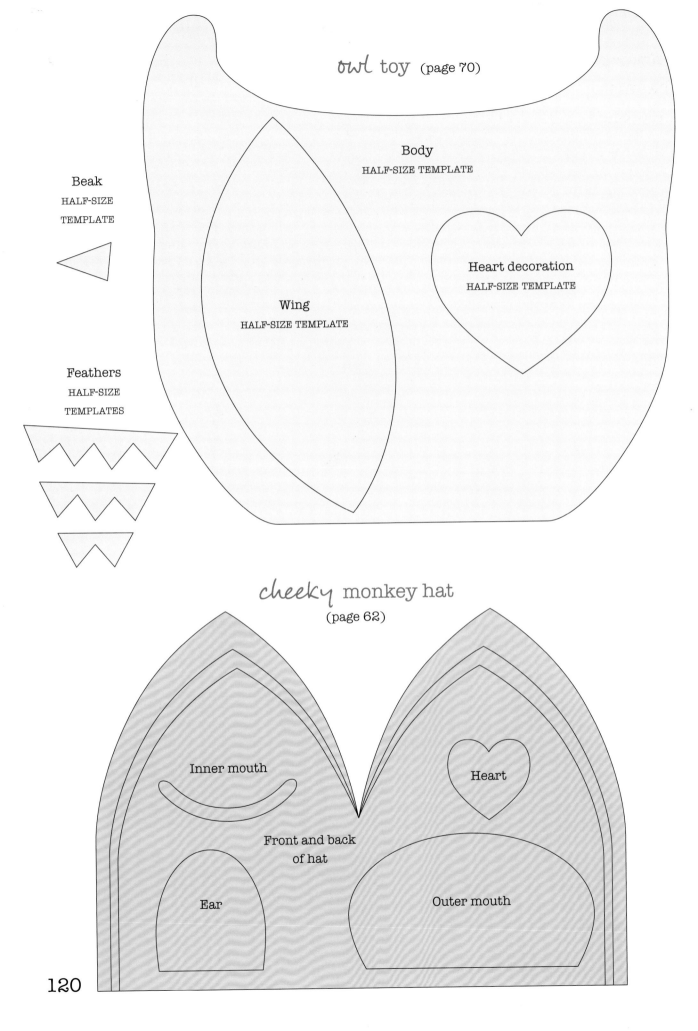

owl toy (page 70)

Body
HALF-SIZE TEMPLATE

Beak
HALF-SIZE
TEMPLATE

Heart decoration
HALF-SIZE TEMPLATE

Wing
HALF-SIZE TEMPLATE

Feathers
HALF-SIZE
TEMPLATES

cheeky monkey hat
(page 62)

Inner mouth

Heart

Front and back
of hat

Ear

Outer mouth

120

pom-pom
beanie
(page 46)

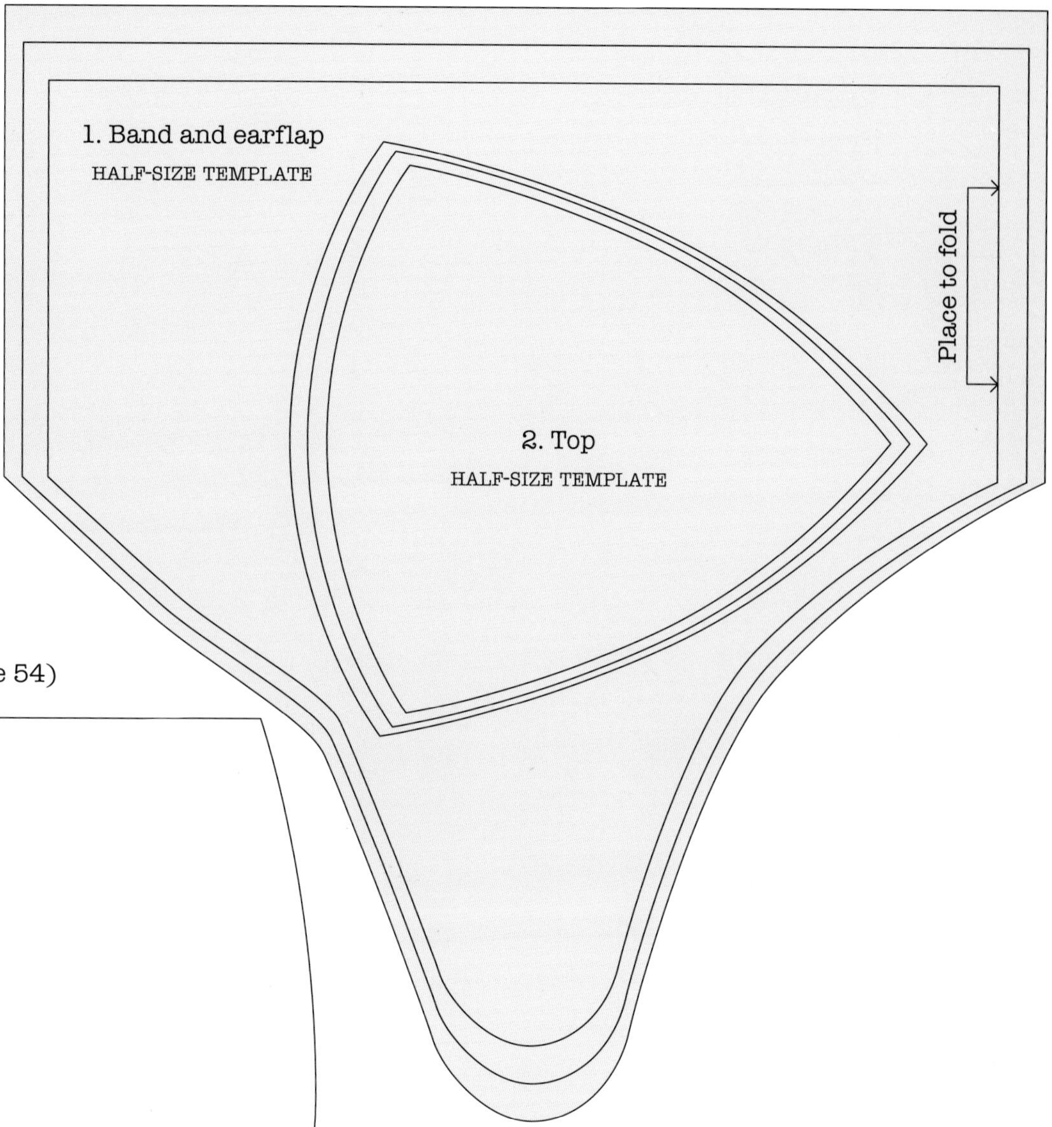

1. Band and earflap
HALF-SIZE TEMPLATE

2. Top
HALF-SIZE TEMPLATE

Place to fold

snake scarf (page 54)

Head
HALF-SIZE TEMPLATE

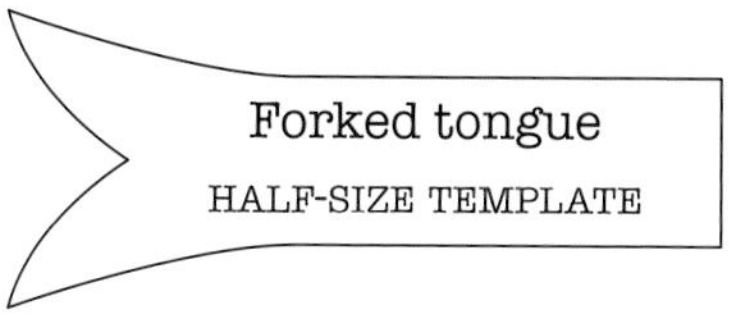

Forked tongue
HALF-SIZE TEMPLATE

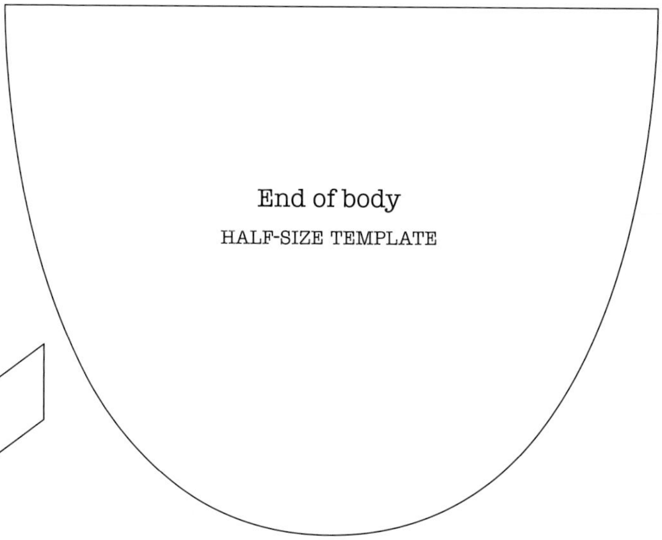

End of body
HALF-SIZE TEMPLATE

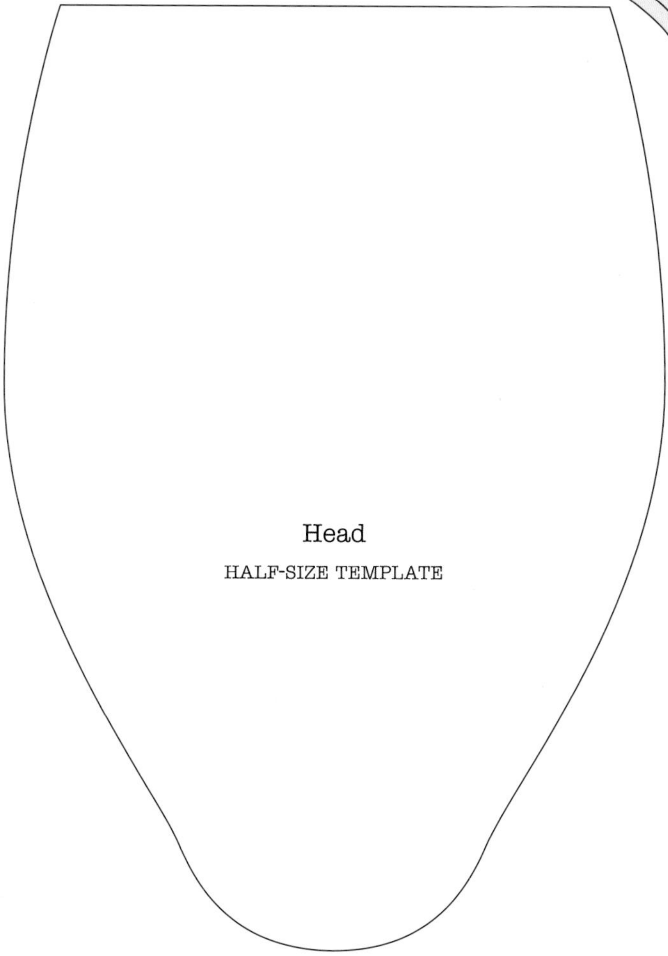

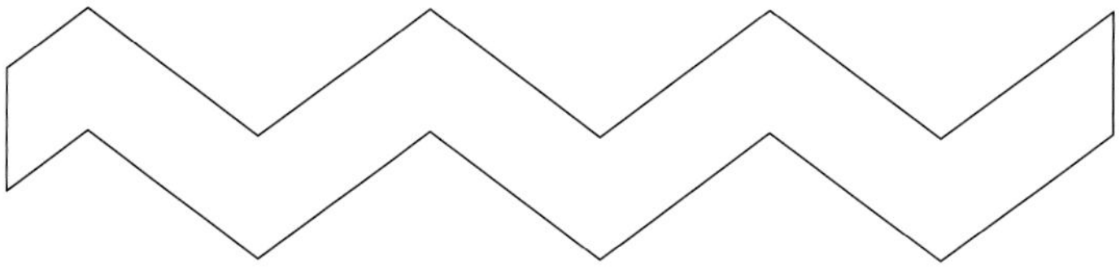

Zig-zag
HALF-SIZE TEMPLATE

frog slippers (page 50)

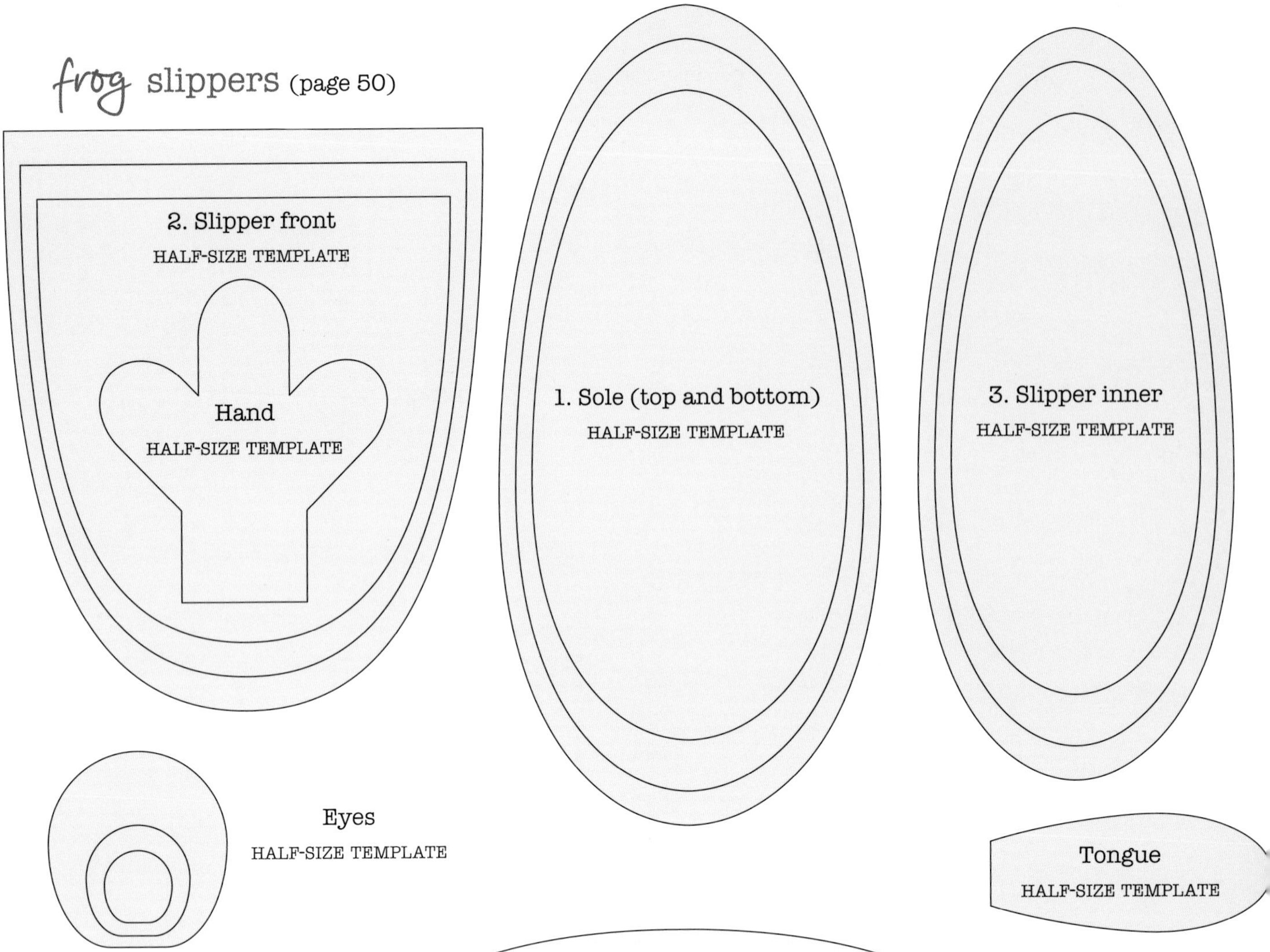

2. Slipper front
HALF-SIZE TEMPLATE

Hand
HALF-SIZE TEMPLATE

1. Sole (top and bottom)
HALF-SIZE TEMPLATE

3. Slipper inner
HALF-SIZE TEMPLATE

Eyes
HALF-SIZE TEMPLATE

Tongue
HALF-SIZE TEMPLATE

kitchen cozies (page 89)

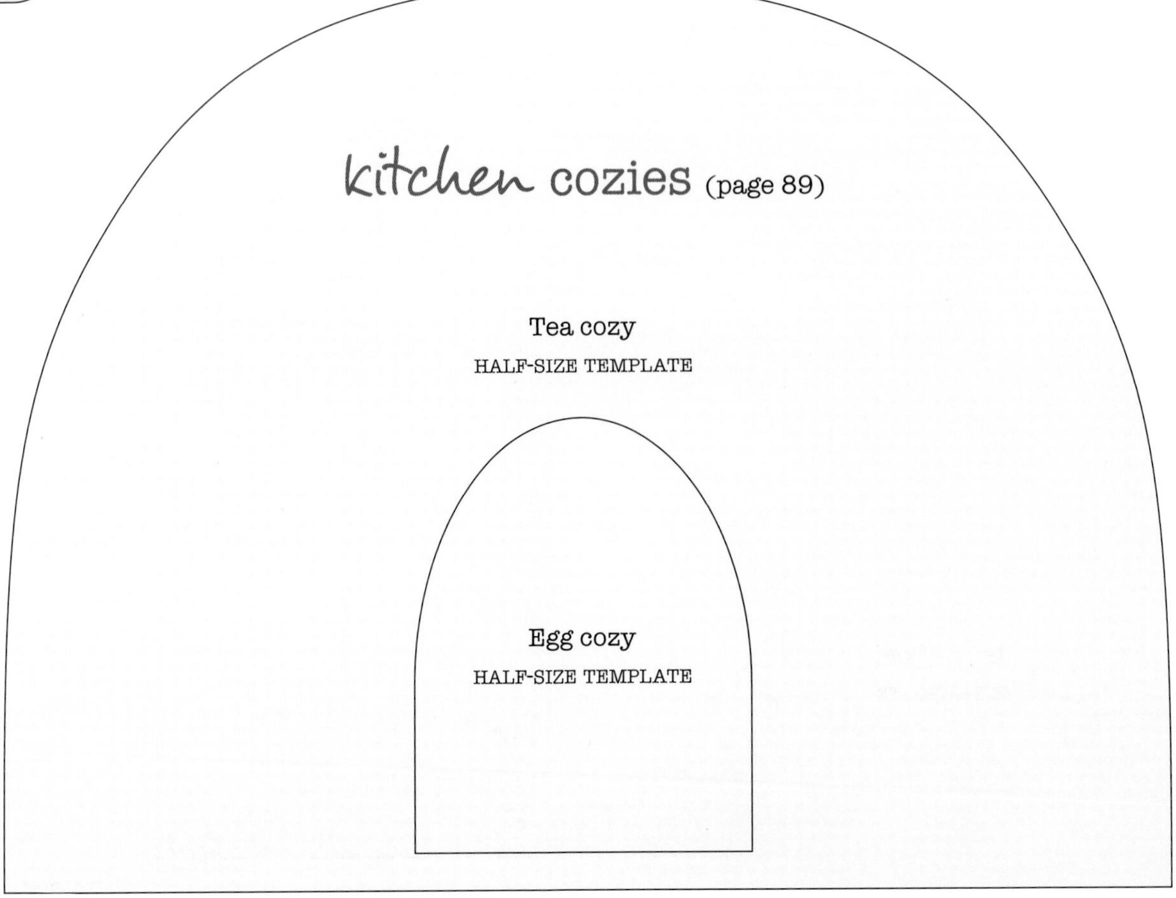

Tea cozy
HALF-SIZE TEMPLATE

Egg cozy
HALF-SIZE TEMPLATE

slipper socks

(page 22)

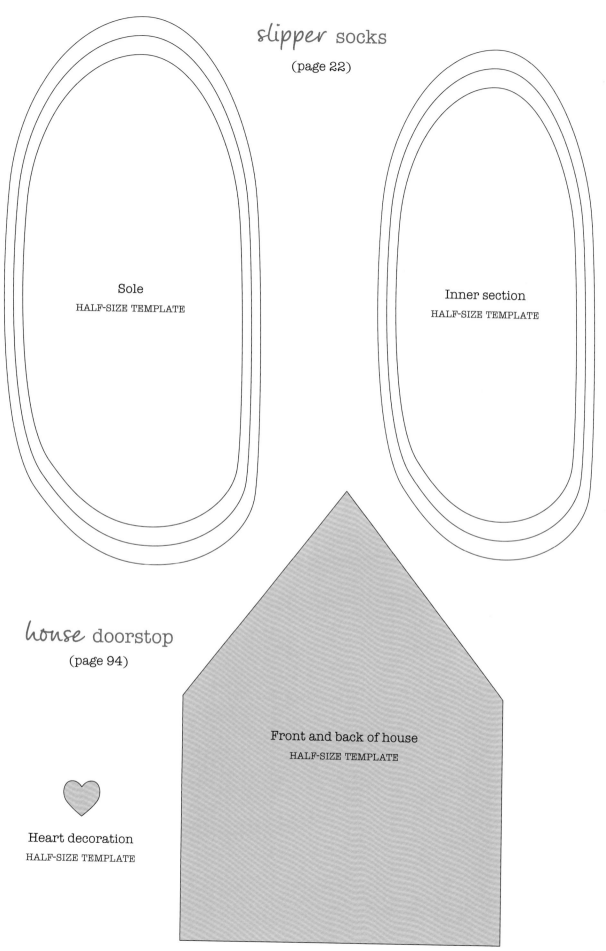

Sole
HALF-SIZE TEMPLATE

Inner section
HALF-SIZE TEMPLATE

house doorstop

(page 94)

Front and back of house
HALF-SIZE TEMPLATE

Heart decoration
HALF-SIZE TEMPLATE

mug cozy (page 86)

retro bottle cozy (page 109)

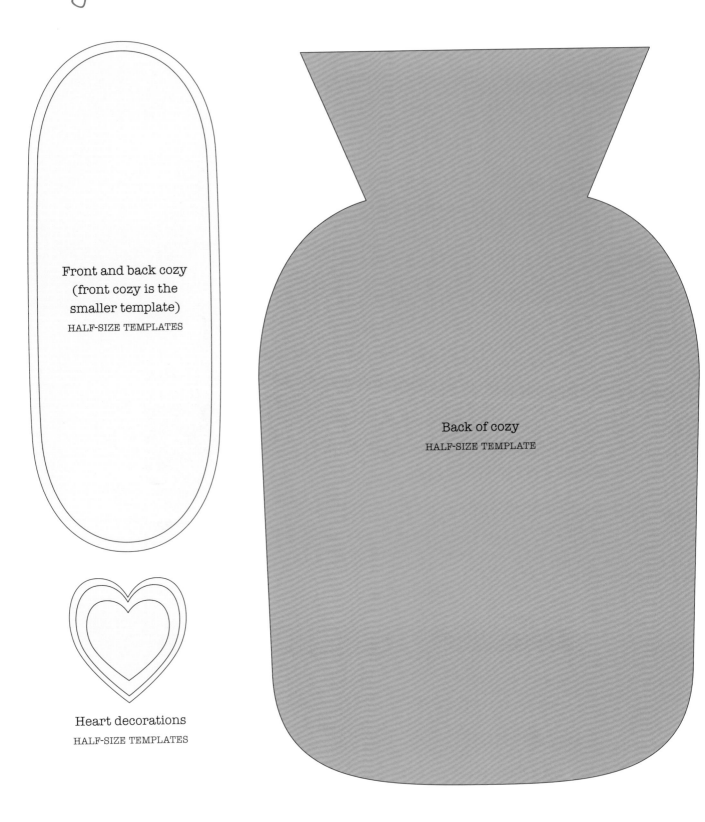

Front and back cozy
(front cozy is the
smaller template)
HALF-SIZE TEMPLATES

Back of cozy
HALF-SIZE TEMPLATE

Heart decorations
HALF-SIZE TEMPLATES

Upper front of cozy

HALF-SIZE TEMPLATE

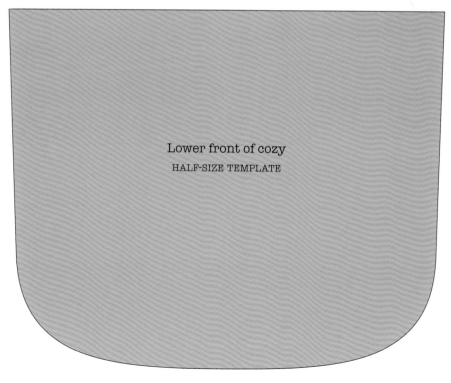

Lower front of cozy

HALF-SIZE TEMPLATE

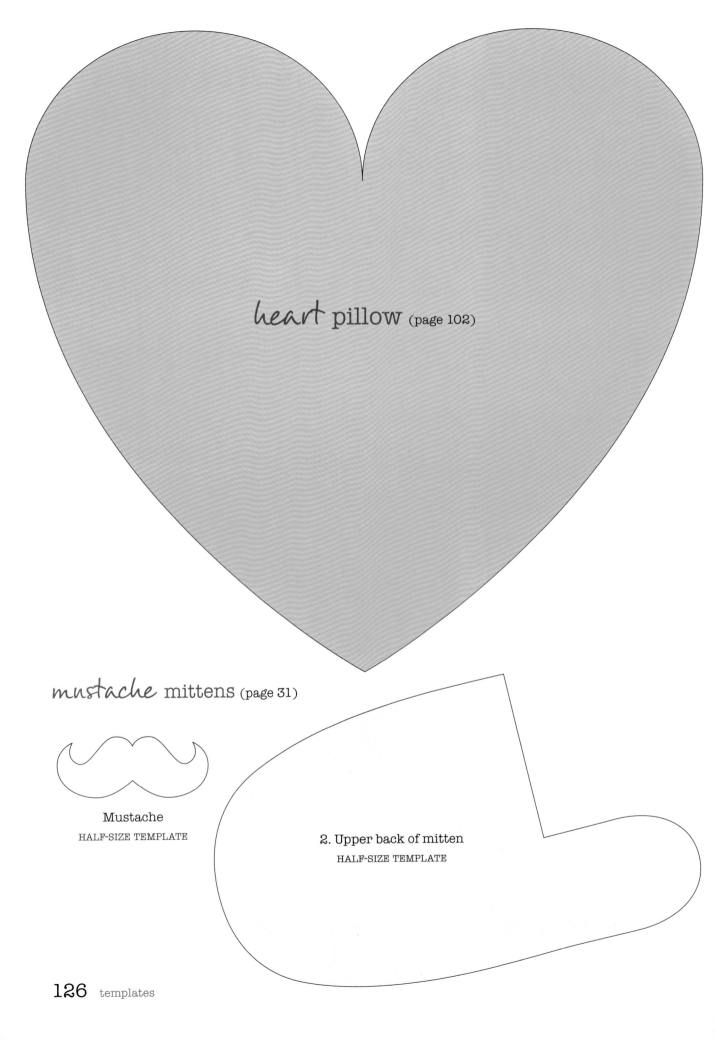

heart pillow (page 102)

mustache mittens (page 31)

Mustache
HALF-SIZE TEMPLATE

2. Upper back of mitten
HALF-SIZE TEMPLATE

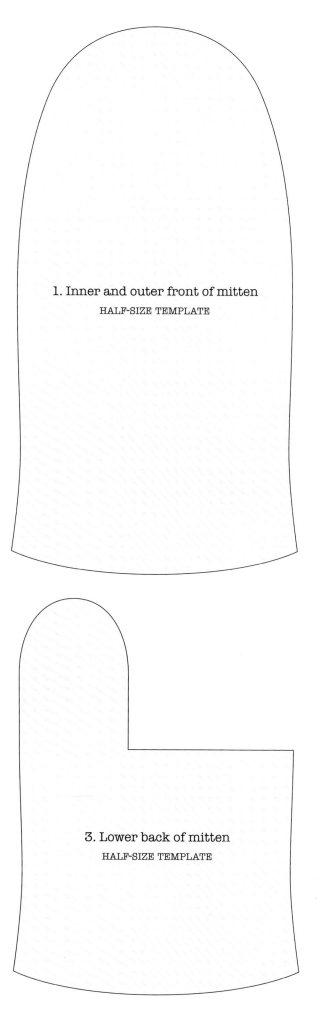

1. Inner and outer front of mitten
HALF-SIZE TEMPLATE

3. Lower back of mitten
HALF-SIZE TEMPLATE

suppliers

Here are some ideas for reliable suppliers. You will be able to use them to source all the materials and tools needed to make the projects in this book.

US

For fleece fabric

Fabric.com www.fabric.com
Fabric Depot www.fabricdepot.com
Jo-Ann Fabric and Craft Stores www.joann.com

For felt, embellishments, and trimmings

Purl Soho www.purlsoho.com
Wool Felt Central www.prairiepointjunction.com

For threads, scissors, embroidery hoops, and other equipment

Hobby Lobby www.hobbylobby.com
Michaels www.michaels.com
A.C. Moore www.acmoore.com

UK

For fleece fabric

Tia Knight www.tiaknightfabrics.co.uk
Plush Addict www.plushaddict.co.uk

For felt, embellishments, and trimmings

Blooming Felt www.bloomingfelt.co.uk
Ebay www.ebay.co.uk
Paper and String www.paper-and-string.co.uk

For threads, scissors, embroidery hoops, and other equipment

Coats Crafts UK www.makeitcoats.com
Hobbycraft www.hobbycraft.co.uk
Homecrafts www.homecrafts.co.uk

index

acknowledgments

Thank you to the CICO team—Carmel Edmonds for commissioning the title, copy editor Sarah Hoggett, and designer Alison Fenton. It's been so lovely to work with you. Thank you to the very talented Michael A Hill for his brilliant illustrations. Thank you to my parents for their never-ending enthusiasm and support. And finally, thank you to my husband Andrew for making my dinner and putting up with me endlessly shouting, "I'm busy sewing!"